Vibrations and Waves

The M.I.T.
Introductory
Physics
Series

Special Relativity
A. P. FRENCH

Vibrations and Waves
A. P. FRENCH

An Introduction to Quantum Physics
A. P. FRENCH and E. F. TAYLOR

The M.I.T.
Introductory
Physics
Series

Vibrations and Waves

A. P. French

PROFESSOR OF PHYSICS, THE MASSACHUSETTS INSTITUTE OF TECHNOLOGY

Van Nostrand Reinhold (UK) Co. Ltd.

Published in 1982 by Van Nostrand Reinhold (UK) Co. Ltd.
Molly Millars Lane, Wokingham, Berkshire, England

Printed in Great Britain by Robert Hartnoll Ltd, Bodmin, Cornwall

Contents

Preface   *ix*

1   Periodic motions   *3*

    Sinusoidal vibrations   4
    The description of simple harmonic motion   5
    The rotating-vector representation   7
    Rotating vectors and complex numbers   10
    Introducing the complex exponential   13
    Using the complex exponential   14
        PROBLEMS   *16*

2   The superposition of periodic motions   *19*

    Superposed vibrations in one dimension   19
    Two superposed vibrations of equal frequency   20
    Superposed vibrations of different frequency; beats   22
    Many superposed vibrations of the same frequency   27
    Combination of two vibrations at right angles   29
    Perpendicular motions with equal frequencies   30
    Perpendicular motions with different frequencies;
      Lissajous figures   35
    Comparison of parallel and perpendicular superposition   38
        PROBLEMS   *39*

3   The free vibrations of physical systems   *41*

    The basic mass–spring problem   41
    Solving the harmonic oscillator equation using complex
      exponentials   43

Elasticity and Young's modulus   45
Floating objects   49
Pendulums   51
Water in a U-tube   53
Torsional oscillations   54
"The spring of air"   57
Oscillations involving massive springs   60
The decay of free vibrations   62
The effects of very large damping   68
    PROBLEMS   70

4   Forced vibrations and resonance   77

Undamped oscillator with harmonic forcing   78
The complex exponential method for forced oscillations   82
Forced oscillations with damping   83
Effect of varying the resistive term   89
Transient phenomena   92
The power absorbed by a driven oscillator   96
Examples of resonance   101
Electrical resonance   102
Optical resonance   105
Nuclear resonance   108
Nuclear magnetic resonance   109
Anharmonic oscillators   110
    PROBLEMS   112

5   Coupled oscillators and normal modes   119

Two coupled pendulums   121
Symmetry considerations   122
The superposition of the normal modes   124
Other examples of coupled oscillators   127
Normal frequencies: general analytical approach   129
Forced vibration and resonance for two coupled oscillators   132
Many coupled oscillators   135
N coupled oscillators   136
Finding the normal modes for N coupled oscillators   139
Properties of the normal modes for N coupled oscillators   141
Longitudinal oscillations   144
N very large   147
Normal modes of a crystal lattice   151
    PROBLEMS   153

6   Normal modes of continuous systems.   Fourier analysis *161*

The free vibrations of stretched strings   162
The superposition of modes on a string   167
Forced harmonic vibration of a stretched string   168

Longitudinal vibrations of a rod   170
The vibrations of air columns   174
The elasticity of a gas   176
A complete spectrum of normal modes   178
Normal modes of a two-dimensional system   181
Normal modes of a three-dimensional system   188
Fourier analysis   189
Fourier analysis in action   191
Normal modes and orthogonal functions   196
    PROBLEMS   197

7   Progressive waves   201

What is a wave?   201
Normal modes and traveling waves   202
Progressive waves in one direction   207
Wave speeds in specific media   209
Superposition   213
Wave pulses   216
Motion of wave pulses of constant shape   223
Superposition of wave pulses   228
Dispersion; phase and group velocities   230
The phenomenon of cut-off   234
The energy in a mechanical wave   237
The transport of energy by a wave   241
Momentum flow and mechanical radiation pressure   243
Waves in two and three dimensions   244
    PROBLEMS   246

8   Boundary effects and interference   253

Reflection of wave pulses   253
Impedances: nonreflecting terminations   259
Longitudinal versus transverse waves: polarization   264
Waves in two dimensions   265
The Huygens–Fresnel principle   267
Reflection and refraction of plane waves   270
Doppler effect and related phenomena   274
Double-slit interference   280
Multiple-slit interference (diffraction grating)   284
Diffraction by a single slit   288
Interference patterns of real slit systems   294
    PROBLEMS   298

A short bibliography   303
Answers to problems   309
Index   313

Preface

THE WORK of the Education Research Center at M.I.T. (formerly the Science Teaching Center) is concerned with curriculum improvement, with the process of instruction and aids thereto, and with the learning process itself, primarily with respect to students at the college or university undergraduate level. The Center was established by M.I.T. in 1960, with the late Professor Francis L. Friedman as its Director. Since 1961 the Center has been supported mainly by the National Science Foundation; generous support has also been received from the Kettering Foundation, the Shell Companies Foundation, the Victoria Foundation, the W. T. Grant Foundation, and the Bing Foundation.

The M.I.T. Introductory Physics Series, a direct outgrowth of the Center's work, is designed to be a set of short books which, taken collectively, span the main areas of basic physics. The series seeks to emphasize the interaction of experiment and intuition in generating physical theories. The books in the series are intended to provide a variety of possible bases for introductory courses, ranging from those which chiefly emphasize classical physics to those which embody a considerable amount of atomic and quantum physics. The various volumes are intended to be compatible in level and style of treatment but are not conceived as a tightly knit package; on the contrary, each book is designed to be reasonably self-contained and usable as an individual component in many different course structures.

The text material in the present volume is intended as an introduction to the study of vibrations and waves in general, but the discussion is almost entirely confined to mechanical systems. Thus, except in a few places, an adequate preparation for it is a good working knowledge of elementary kinematics and dynamics. The decision to limit the scope of the book in this way was guided by the fact that the presentation is quantitative and analytical rather than descriptive. The temptation to incorporate discussions of electrical and optical systems was always strong, but it was felt that a great part of the language of the subject could be developed most simply and straightforwardly in terms of mechanical displacements and scalar wave equations, with only an occasional allusion to other systems.

On the matter of mathematical background, a fair familiarity with calculus is assumed, such that the student will recognize the statement of Newton's law for a harmonic oscillator as a differential equation and be readily able to verify its solution in terms of sinusoidal functions. The use of the complex exponential for the analysis of oscillatory systems is introduced at an early stage; the necessary introduction of partial differential equations is, however, deferred until fairly late in the book. Some previous experience with a calculus course in which differential equations have been discussed is certainly desirable, although it is not in the author's view essential.

The presentation lays more emphasis on the concept of normal modes than is customary in introductory courses. It is the author's belief, as stated in the text, that this can greatly enrich the student's understanding of how the dynamics of a continuum can be linked to the dynamics of one or a few particles. What is not said, but has also been very much in mind, is that the development and use of such features as orthogonality and completeness of a set of normal modes will give to the student a sense of old acquaintance renewed when he meets these features again in the context of quantum mechanics.

Although the emphasis is on an analytical approach, the effort has been made to link the theory to real examples of the phenomena, illustrated where possible with original data and photographs. It is intended that this "documentation" of the subject should be a feature of all the books in the series.

This book, like the others in the series, owes much to the thoughts, criticisms, and suggestions of many people, both students and instructors. A special acknowledgment is due to

Prof. Jack R. Tessman (Tufts University), who was deeply involved with our earliest work on this introductory physics program and who, with the present author, taught a first trial version of some of the material at M.I.T. during 1963–1964. Much of the subsequent writing and rewriting was discussed with him in detail. In particular, in the present volume, the introduction to coupled oscillators and normal modes in Chapter 5 stems largely from the approach that he used in class.

Thanks are due to the staff of the Education Research Center for help in the preparation of this volume, with special mention of Miss Martha Ransohoff for her enthusiastic efforts in typing the final manuscript and to Jon Rosenfeld for his work in setting up and photographing a number of demonstrations for the figures.

A. P. FRENCH

Cambridge, Massachusetts
July 1970

Vibrations and waves

*These are the Phenomena of Springs and springy bodies,
which as they have not hitherto been by any that I know
reduced to Rules, so have all the attempts for the
explications of the reason of their power, and of springiness
in general, been very insufficient.*

ROBERT HOOKE, *De Potentia Restitutiva* (1678)

1
Periodic motions

THE VIBRATIONS or oscillations of mechanical systems constitute one of the most important fields of study in all physics. Virtually every system possesses the capability for vibration, and most systems can vibrate freely in a large variety of ways. Broadly speaking, the predominant natural vibrations of small objects are likely to be rapid, and those of large objects are likely to be slow. A mosquito's wings, for example, vibrate hundreds of times per second and produce an audible note. The whole earth, after being jolted by an earthquake, may continue to vibrate at the rate of about one oscillation per hour. The human body itself is a treasure-house of vibratory phenomena; as one writer has put it[1]:

> After all, our hearts beat, our lungs oscillate, we shiver when we are cold, we sometimes snore, we can hear and speak because our eardrums and larynges vibrate. The light waves which permit us to see entail vibration. We move by oscillating our legs. We cannot even say "vibration" properly without the tip of the tongue oscillating ... Even the atoms of which we are constituted vibrate.

The feature that all such phenomena have in common is *periodicity*. There is a pattern of movement or displacement that repeats itself over and over again. This pattern may be simple

[1]From R. E. D. Bishop, *Vibration*, Cambridge University Press, New York, 1965. A most lively and fascinating general account of vibrations with particular reference to engineering problems.

3

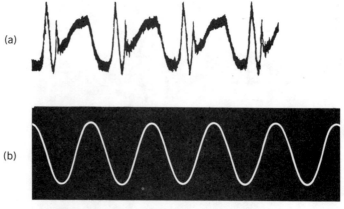

(a)

(b)

Fig. 1-1 (a) Pressure variations inside the heart of a
cat (After Straub, in E. H. Starling, Elements of
Human Physiology, Churchill, London, 1907.)
(b) Vibrations of a tuning fork.

or complicated; Fig. 1-1 shows an example of each—the rather
complex cycle of pressure variations inside the heart of a cat, and
the almost pure sine curve of the vibrations of a tuning fork. In
each case the horizontal axis represents the steady advance of
time, and we can identify the length of time—the period T—
within which one complete cycle of the vibration is performed.

In this book we shall study a number of aspects of periodic
motions, and will proceed from there to the closely related phenomenon of progressive waves. We shall begin with some discussion of the purely kinematic description of vibrations. Later,
we shall go into some of the dynamical properties of vibrating
systems—those dynamical features that allow us to see oscillatory
motion as a real physical problem, not just as a mathematical
exercise.

SINUSOIDAL VIBRATIONS

Our attention will be directed overwhelmingly to sinusoidal
vibrations of the sort exemplified by Fig. 1-1(b). There are two
reasons for this—one physical, one mathematical, and both basic
to the whole subject. The physical reason is that purely sinusoidal
vibrations do, in fact, arise in an immense variety of mechanical
systems, being due to restoring forces that are proportional to
the displacement from equilibrium. Such motion is almost always
possible if the displacements are small enough. If, for example,
we have a body attached to a spring, the force exerted on it at a

displacement x from equilibrium may be written

$$F(x) = -(k_1 x + k_2 x^2 + k_3 x^3 + \cdots)$$

where k_1, k_2, k_3, etc., are a set of constants, and we can always find a range of values of x within which the sum of the terms in x^2, x^3, etc., is negligible, according to some stated criterion (e.g., 1 part in 10^3, or 1 part in 10^6) compared to the term $-k_1 x$, unless k_1 itself is zero. If the body is of mass m and the mass of the spring is negligible, the equation of motion of the body then becomes

$$m \frac{d^2 x}{dt^2} = -k_1 x$$

which, as one can readily verify, is satisfied by an equation of the form

$$x = A \sin(\omega t + \varphi_0) \tag{1-1}$$

where $\omega = (k_1/m)^{1/2}$. This brief discussion will be allowed to serve as a reminder that sinusoidal vibration—simple harmonic motion—is a prominent possibility in small vibrations, but also that in general it is only an approximation (although perhaps a very close one) to the true motion.

The second reason—the mathematical one—for the profound importance of purely sinusoidal vibrations is to be found in a famous theorem propounded by the French mathematician J. B. Fourier in 1807. According to Fourier's theorem, *any* disturbance that repeats itself regularly with a period T can be built up from (or is analyzable into) a set of pure sinusoidal vibrations of periods T, $T/2$, $T/3$, etc., with appropriately chosen amplitudes— i.e., an infinite series made up (to use musical terminology) of a fundamental frequency and all its harmonics. We shall have more to say about this later, but we draw attention to Fourier's theorem at the outset so as to make it clear that we are not limiting the scope or applicability of our discussions by concentrating on simple harmonic motion. On the contrary, a thorough familiarity with sinusoidal vibrations will open the door to every conceivable problem involving periodic phenomena.

THE DESCRIPTION OF SIMPLE HARMONIC MOTION

A motion of the type described by Eq. (1-1), simple harmonic motion (SHM),[1] is represented by an $x - t$ graph such as that

[1] This convenient and widely used abbreviation is one that we shall employ often.

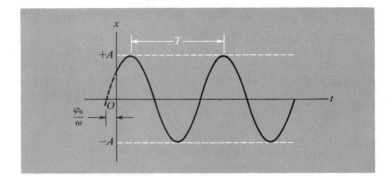

Fig. 1-2 Simple
harmonic motion of
period T and
amplitude A.

shown in Fig. 1–2. We recognize the characteristic features of
any such sinusoidal disturbance:

1. It is confined within the limits $x = \pm A$. The positive
quantity A is the *amplitude* of the motion.

2. The motion has the period T equal to the time between
successive maxima, or more generally between successive occa-
sions on which both the displacement x and the velocity dx/dt
repeat themselves. Given the basic equation (1–1),

$$x = A \sin(\omega t + \varphi_0)$$

the period must correspond to an increase by the amount 2π in
the argument of the sine function. Thus we have

$$\omega(t + T) + \varphi_0 = (\omega t + \varphi_0) + 2\pi$$

whence

$$T = \frac{2\pi}{\omega} \tag{1-2}$$

The situation at $t = 0$ (or at any other designated time, for that
matter) is completely specified if one states the values of both x
and dx/dt at that instant. For the particular time $t = 0$, let
these quantities be denoted by x_0 and v_0, respectively. Then we
have the following identities:

$$x_0 = A \sin \varphi_0$$
$$v_0 = \omega A \cos \varphi_0$$

If the motion is known to be described by an equation of the
form (1–1), these last two relationships can be used to calculate
the amplitude A and the angle φ_0 (the initial phase angle of
the motion):

$$A = \left[x_0{}^2 + \left(\frac{v_0}{\omega}\right)^2 \right]^{1/2} \qquad \varphi_0 = \tan^{-1}\left(\frac{\omega x_0}{v_0}\right)$$

6   Periodic motions

The value of the angular frequency ω of the motion is here assumed to be independently known.

Equation (1–1) as it stands defines a sinusoidal variation of x with t over the whole range of t, regarded as a purely mathematical variable, from $-\infty$ to $+\infty$. Since every real vibration has a beginning and an end, it cannot therefore, even if purely sinusoidal while it lasts, be properly described by Eq. (1–1) alone. If, for example, a simple harmonic vibration were started at $t = t_1$ and stopped at $t = t_2$, its complete description in mathematical terms would require a total of three statements:

$$-\infty < t < t_1 \qquad x = 0$$
$$t_1 \leq t \leq t_2 \qquad x = A \sin(\omega t + \varphi_0)$$
$$t_2 < t < \infty \qquad x = 0$$

This limitation on the validity of Eq. (1–1) as a complete description of a physically real harmonic vibration should always be borne in mind. It is not just a mathematical quibble. As judged by strictly *physical* criteria, a vibration does not appear to be effectively a pure sinusoid unless it continues for a very large number of periods. For example, if the ear were allowed to receive only one complete cycle of the sound from a tuning fork, vibrating as in Fig. 1–1(b), the aural impression would not at all be that of a pure tone at the characteristic frequency of the fork, but would instead be a confused jangle of tones.[1] It would be premature, and in a sense irrelevant, to discuss the phenomenon in any more detail at this point; the problem is again one of Fourier analysis. What is important at this stage is to recognize that the simple harmonic vibrations of an actual physical system must be long-continued—must represent what is often called a steady state of vibration—for Eq. (1–1) by itself to be used as an acceptable description of them.

THE ROTATING-VECTOR REPRESENTATION

One of the most useful ways of describing simple harmonic motion is obtained by regarding it as the projection of uniform

[1]The complexity of the sound could be more convincingly demonstrated with an automatic wave analyzer, because it is known that what we hear is not an exact replica of an incoming sound wave—the ear adds distortions of its own. See, for example, W. A. Van Bergeijk, J. R. Pierce, and E. A. David, *Waves and the Ear*, Doubleday (Anchor Book), New York, 1960.

circular motion. Imagine, for example, that a disk of radius A rotates about a vertical axis at the rate of ω rad/sec. Suppose that a peg P is attached to the edge of the disk and that a horizontal beam of parallel light casts a shadow of the peg on a vertical screen, as shown in Fig. 1–3(a). Then this shadow performs simple harmonic motion with period $2\pi/\omega$ and amplitude A along a horizontal line on the screen.

More abstractly, we can imagine SHM as being the *geometrical* projection of uniform circular motion. (By geometrical projection we mean simply the process of drawing a perpendicular to a given line from the instantaneous position of the point P.)

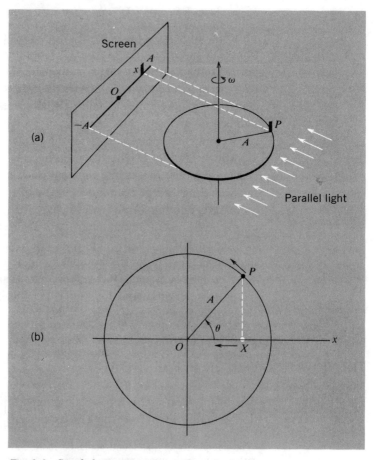

Fig. 1–3 Simple harmonic motion as the projection in its own plane of uniform circular motion.

In Fig. 1–3(b) we indicate the way in which the end point of the rotating vector OP can be projected onto a diameter of the circle. In particular we choose the horizontal axis Ox as the line along which the actual oscillation takes place. The instantaneous position of the point P is then defined by the constant length A and the variable angle θ. It will be in accord with our usual conventions for polar coordinates if we take the counterclockwise direction as positive; the actual value of θ can be written

$$\theta = \omega t + \alpha$$

where α is the value of θ at $t = 0$.

As specified above, the displacement x of the actual motion is given by

$$x = A \cos \theta = A \cos(\omega t + \alpha) \tag{1-3}$$

Superficially, this equation differs from our initial description of simple harmonic motion according to Eq. (1–1). We can, however, readily satisfy the requirement that they be identical, because for any angle θ we have

$$\cos \theta = \sin \left(\theta + \frac{\pi}{2} \right)$$

The identity of Eqs. (1–1) and (1–3) requires

$$A \sin(\omega t + \varphi_0) = A \cos(\omega t + \alpha)$$

i.e.,

$$\sin(\omega t + \varphi_0) = \sin \left(\omega t + \alpha + \frac{\pi}{2} \right)$$

The sines of two angles are equal if the angles are equal or if they differ by any integral multiple of 2π. Taking the simplest of these possibilities, we can thus put

$$\varphi_0 = \alpha + \frac{\pi}{2} \tag{1-4}$$

The equivalence of Eqs. (1–1) and (1–3) subject to the above condition allows us to describe any simple harmonic vibration equally well in terms of a sine or a cosine function. In much of our future analysis, however, it will prove to be extremely profitable to fix upon the cosine form, so as to exploit the description of the displacement as the projection of a uniformly rotating vector on the reference axis of plane polar coordinates. The use of this approach in all its richness hinges upon some mathematical ideas which will be the subject of the next sections.

9 The rotating-vector representation

The use of a uniform circular motion as a purely geometrical basis for describing SHM embodies more than we have so far chosen to recognize. This circular motion, once we have set it up, defines SHM of amplitude A and angular frequency ω along *any* straight line in the plane of the circle. In particular, if we imagine a y axis perpendicular to the real physical axis Ox of the actual motion, the rotating vector OP defines for us, in addition to the true oscillation along x, an accompanying orthogonal oscillation along y, such that

$$x = A \cos(\omega t + \alpha)$$
$$y = A \sin(\omega t + \alpha)$$

$(1-5)$

And even though this motion along y has no actual existence, we can proceed precisely *as if* we were dealing always with the motion of a point in two dimensions, as described by equations $(1-5)$, provided that, at the end, we extract only the x component, because this is the physically meaningful result of the motion thus described.

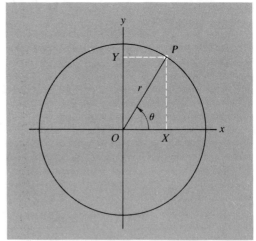

Fig. 1–4 Cartesian and polar representations of a rotating vector.

There exists an unambiguous way of establishing and maintaining the distinction between the physically real and the physically unreal components of the motion. Suppose that a vector OP (Fig. 1–4) has the plane polar coordinates (r, θ). The rectangular (Cartesian) components (x, y) are, of course, defined by the following equations:

$$x = r \cos \theta \qquad y = r \sin \theta$$

The complete vector **r** can then be expressed as the vector sum of these two orthogonal components. If we chose to employ the customary notation of vector analysis, we would introduce a unit vector **i** to denote displacement along x, and a unit vector **j** to denote displacement along y. We should then put

$$\mathbf{r} = \mathbf{i}x + \mathbf{j}y$$

But without any sacrifice of informational content, we can define the vector by means of the following equation:

$$\mathbf{r} = x + jy \tag{1-6}$$

All that is required is an initial convention by which it is agreed that Eq. (1–6) embodies the following statements:

  1. A displacement, such as x, without any qualifying factors, is to be made in a direction parallel to the x axis.
  2. The term jy is to be read as an instruction to make the displacement y in a direction parallel to the y axis. It is, in fact, customary to dispense with the usual vector symbolism altogether, by introducing a quantity z, understood to be the result of adding jy to x—i.e., identical with **r** as defined above. Thus we put

$$z = x + jy \tag{1-7}$$

We now proceed to broaden the interpretation of the symbol j, by reading it as an instruction to perform a counterclockwise rotation of 90° upon whatever it precedes. Consider the following specific examples:

  a. To form the quantity jb, we step off a distance b along the x axis and then rotate through 90° so as to end up with a displacement of length b along y.
  b. To form the quantity j^2b we first form jb, as above, and then apply to it a further 90° rotation—i.e., we identify j^2b as $j(jb)$. But this at once leads to an important identity. Two successive 90° rotations in the same sense convert a displacement b (along the positive x direction) into the displacement $-b$. Hence we set up the algebraic identity

$$j^2 = -1 \tag{1-8}$$

The quantity j itself can thus be regarded, algebraically speaking, as a square root of -1. (And $-j$ is another square root, also satisfying the above equation.)[1]

[1] The use of the symbol j for $\sqrt{-1}$ has emerged rather naturally from our quasi-geometrical approach. Very often, however, in mathematics texts, one will find the symbol i used for this purpose. Physicists and engineers tend to

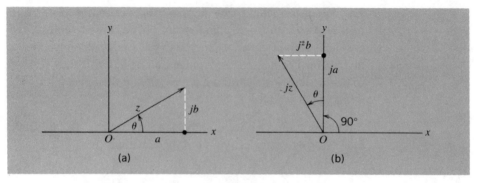

Fig. 1–5 (a) Representation of a vector in the complex plane.
(b) Multiplication of z by j is equivalent to a 90° rotation.

c. Suppose we take a vector z having an x component of length a and a y component of length b (Fig. 1–5a). What is jz? We have

$$z = a + jb$$
$$jz = ja + j^2b$$
$$= ja + (-b)$$

The summation of the new vector components on the right of the above equation is shown in Fig. 1–5(b). The recipe is consistent! The resultant vector jz is obtained from the original vector z just by the extra rotation of 90°.

Whether or not you have been introduced to this kind of analysis previously, you will be able to recognize that we are walking along a dividing line—or, more properly, a bridge—between geometry and algebra. If the quantities a and b are real numbers, as we have assumed in example c, then the combination $z = a + jb$ is what is known as a complex number. But in geometrical terms it can be regarded as a displacement along an axis at some angle θ to the x axis, such that $\tan \theta = b/a$, as is clear from Fig. 1–5(a).

In this representation of a vector by a complex number, we have an automatic way of selecting out the physically relevant part for the purpose of analyzing simple harmonic motion. If, after solving an oscillatory motion problem in these terms, we obtain a final answer in the form $z = a + jb$, where a and b are both real numbers, then the quantity a is the wanted quantity, and b can be discarded.

prefer the j notation, so as to reserve the symbol i for electric current—a not insignificant consideration because the mathematical techniques we are developing here find some of their most important uses in connection with electrical circuit problems.

A quantity of the form jb alone (with b real) is called purely imaginary. From the standpoint of mathematics as such, this is perhaps an unfortunate term, because in the extension of the concept of number from real to complex an "imaginary" component such as jb is on an equal footing with a real component such as a. But as applied to the analysis of one-dimensional oscillations, this terminology conforms perfectly, as we have already seen, to the physically real and unreal parts of an imagined two-dimensional motion.

INTRODUCING THE COMPLEX EXPONENTIAL

The preceding discussion may not seem to have added much to our earlier analysis. But now we are ready for the chief character, the mathematical function toward which this development has been directed. This is the complex exponential function—or, to be more specific, the exponential function in the case in which the exponent is imaginary in the mathematical sense mentioned at the end of the last section. After introducing this function, we shall find that our efforts in doing so are repaid many times over in terms of the ease of handling oscillatory problems. Not all of these benefits will be apparent right away, but they will come to be appreciated more and more as one digs deeper into the subject.

We begin by taking the series expansions of the sine and cosine functions:

$$\sin \theta = \theta - \frac{\theta^3}{3!} + \frac{\theta^5}{5!} \cdots \tag{1-9}$$

$$\cos \theta = 1 - \frac{\theta^2}{2!} + \frac{\theta^4}{4!} \cdots \tag{1-10}$$

These expansions, if not already familiar, are readily developed with the help of Taylor's theorem.[1]

Let us now form the following combination:

$$\cos \theta + j \sin \theta = 1 + j\theta - \frac{\theta^2}{2!} - j\frac{\theta^3}{3!} + \frac{\theta^4}{4!} + \cdots \tag{1-11}$$

---

[1]By Taylor's theorem,

$$f(x) = f(0) + xf'(0) + \frac{x^2}{2!}f''(0) + \cdots$$

Therefore,

$$\sin \theta = \sin 0 + \theta \cos 0 + \frac{\theta^2}{2!}(-\sin 0) + \frac{\theta^3}{3!}(\cos 0) \cdots$$

$$\cos \theta = \cos 0 + \theta(-\sin 0) + \frac{\theta^2}{2!}(-\cos 0) + \frac{\theta^3}{3!}(\sin 0) \cdots$$

We have seen that -1 is expressible as j^2, so the above equation can be rewritten as follows:

$$\cos \theta + j \sin \theta = 1 + j\theta + \frac{(j\theta)^2}{2!} + \frac{(j\theta)^3}{3!} + \cdots$$
$$+ \frac{(j\theta)^n}{n!} + \cdots \qquad (1\text{–}12)$$

But the right-hand side of this equation has precisely the form of the exponential series, with the exponent set equal to $j\theta$. Thus we are enabled to write the following identity:

$$\cos \theta + j \sin \theta = e^{j\theta} \qquad (1\text{–}13)$$

This is a very dramatic result, mathematically speaking, providing a clear connection between plane geometry (as represented by trigonometric functions) and algebra (as represented by the exponential function). R. P. Feynman has called it "this amazing jewel ... the most remarkable formula in mathematics."[1] It was set up by Leonhard Euler in 1748.

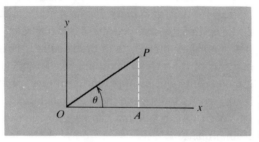

Fig. 1–6 Geometrical interpretation of Euler's relation, $e^{j\theta} = \cos \theta + j \sin \theta$.

Let us display the geometrical character of the result. Using "real" and "imaginary" axes Ox, Oy (Fig. 1–6) we draw OA of length equal to $\cos \theta$, and AP of length equal to $\sin \theta$. The vector sum of these is OP; it is clearly *of length unity* and it makes the angle θ with the x axis. More generally, the multiplication of any complex number z by $e^{j\theta}$ is describable, in geometrical terms, as a positive rotation, through the angle θ, of the vector by which z may be represented—without any alteration of its length. (*Exercise*: Verify this.)

USING THE COMPLEX EXPONENTIAL

Why should the introduction of Eq. (1–13) be such an important

[1] R. P. Feynman, R. B. Leighton, and M. L. Sands, *Feynman Lectures on Physics*, Vol. I, Addison-Wesley, Reading, Mass., 1963.

contribution to the analysis of vibrations? The prime reason is the special property of the exponential function—its reappearance after every operation of differentiation or integration. For the problems that we shall be concerned with are problems involving periodic displacements and the time derivatives of these displacements. If, as often happens, the basic equation of motion contains terms proportional to velocity and acceleration, as well as to displacement itself, then the use of a simple trigonometric function to describe the motion leads to an awkward mixture of sine and cosine terms. For example:

If

$$x = A \cos(\omega t + \alpha)$$

then

$$\frac{dx}{dt} = -\omega A \sin(\omega t + \alpha)$$

$$\frac{d^2 x}{dt^2} = -\omega^2 A \cos(\omega t + \alpha)$$

On the other hand, if we work with the combination $x + jy$, with x and y as given by equations (1–5), we have the following:

$$z = A \cos(\omega t + \alpha) + jA \sin(\omega t + \alpha)$$

i.e.,

$$z = Ae^{j(\omega t + \alpha)}$$

with

$$x = \text{real part of } z[1]$$

Then

$$\frac{dz}{dt} = j\omega Ae^{j(\omega t + \alpha)} = j\omega z$$

$$\frac{d^2 z}{dt^2} = (j\omega)^2 Ae^{j(\omega t + \alpha)} = -\omega^2 z$$

These three vectors are shown in Fig. 1–7 (using three separate diagrams, because quantities of three physically different kinds—displacement, velocity, acceleration—are being described). In each case the physically relevant component is recognizable as being the real component of the vector in question, and the phase relationships are visible at a glance (given the result that each factor of j is to be read as an advance in phase angle by $\pi/2$). This is a very trivial example that does not really display the

[1]Often abbreviated Re(z).

15   Using the complex exponential

Fig. 1-7 (a) Displacement vector z and its real projection x. (b) Velocity vector dz/dt and its real projection dx/dt. (c) Acceleration vector d^2z/dt^2 and its real projection d^2x/dt^2.

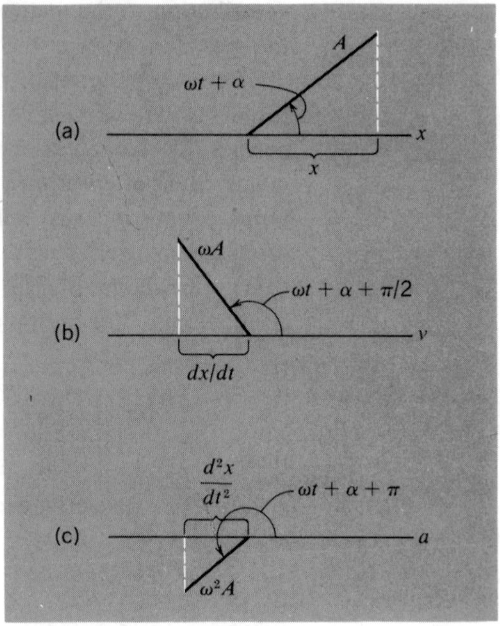

power of the method, but we shall come to some more substantial applications quite shortly.

PROBLEMS

1-1 Consider a vector z defined by the equation $z = z_1 z_2$, where $z_1 = a + jb$, $z_2 = c + jd$.

(a) Show that the length of z is the product of the lengths of z_1 and z_2.

(b) Show that the angle between z and the x axis is the sum of the angles made by z_1 and z_2 separately.

1-2 Consider a vector z defined by the equation $z = z_1/z_2$ $(z_2 \neq 0)$, where $z_1 = a + jb$, $z_2 = c + jd$.

(a) Show that the length of z is the quotient of the lengths of z_1 and z_2.

(b) Show that the angle between z and the x axis is the difference of the angles made by z_1 and z_2 separately.

1-3 Show that the multiplication of any complex number z by $e^{i\theta}$ is describable, in geometrical terms, as a positive rotation through the angle θ of the vector by which z is represented, without any alteration of its length.

1-4 (a) If $z = Ae^{i\theta}$, deduce that $dz = jz \, d\theta$, and explain the meaning of this relation in a vector diagram.

(b) Find the magnitudes and directions of the vectors $(2 + j\sqrt{3})$ and $(2 - j\sqrt{3})^2$.

16 Periodic motions

1-5 To take successive derivatives of $e^{j\theta}$ with respect to θ, one merely multiplies by j:

$$\frac{d}{d\theta}(Ae^{j\theta}) = jAe^{j\theta}$$

Show that this prescription works if the sinusoidal representation $e^{j\theta} = \cos\theta + j\sin\theta$ is used.

1-6 Given Euler's relation $e^{j\theta} = \cos\theta + j\sin\theta$, find
   (a) The geometric representation of $e^{-j\theta}$.
   (b) The exponential representation of $\cos\theta$.
   (c) The exponential representation of $\sin\theta$.

1-7 (a) Justify the formulas $\cos\theta = (e^{j\theta} + e^{-j\theta})/2$ and $\sin\theta = (e^{j\theta} - e^{-j\theta})/2j$, using the appropriate series.
   (b) Display the above relationships geometrically by means of vector diagrams in the xy plane.

1-8 Using the exponential representations for $\sin\theta$ and $\cos\theta$, verify the following trigonometric identities:
   (a) $\sin^2\theta + \cos^2\theta = 1$ (b) $\cos^2\theta - \sin^2\theta = \cos 2\theta$
   (c) $2\sin\theta\cos\theta = \sin 2\theta$

1-9 Would you be willing to pay 20 cents for an object valued by a mathematician at $\$j^j$? (Remember that $\cos\theta + j\sin\theta = e^{j\theta}$.)

1-10 Verify that the differential equation $d^2y/dx^2 = -ky$ has as its solution

$$y = A\cos(kx) + B\sin(kx)$$

where A and B are arbitrary constants. Show also that this solution can be written in the form

$$y = C\cos(kx + \alpha) = C\,\mathrm{Re}[e^{j(kx+\alpha)}] = \mathrm{Re}[(Ce^{j\alpha})e^{jkx}]$$

and express C and α as functions of A and B.

1-11 A mass on the end of a spring oscillates with an amplitude of 5 cm at a frequency of 1 Hz (cycles per second). At $t = 0$ the mass is at its equilibrium position ($x = 0$).
   (a) Find the possible equations describing the position of the mass as a function of time, in the form $x = A\cos(\omega t + \alpha)$, giving the numerical values of A, ω, and α.
   (b) What are the values of x, dx/dt, and d^2x/dt^2 at $t = \frac{8}{3}$ sec?

1-12 A point moves in a circle at a constant speed of 50 cm/sec. The period of one complete journey around the circle is 6 sec. At $t = 0$ the line to the point from the center of the circle makes an angle of $30°$ with the x axis.
   (a) Obtain the equation of the x coordinate of the point as a function of time, in the form $x = A\cos(\omega t + \alpha)$, giving the numerical values of A, ω, and α.
   (b) Find the values of x, dx/dt, and d^2x/dt^2 at $t = 2$ sec.

17   Problems

"... *That undulation, each way free—*
It taketh me."

MICHAEL BARSLEY (1937), *On his Julia, walking*
(After Robert Herrick)

2
The superposition
of
periodic motions

SUPERPOSED VIBRATIONS IN ONE DIMENSION

MANY PHYSICAL situations involve the simultaneous application of two or more harmonic vibrations to the same system. Examples of this are especially common in acoustics. A phonograph stylus, a microphone diaphragm, or a human eardrum is in general being subjected to a complicated combination of such vibrations, resulting in some over-all pattern of its displacement as a function of time. We shall consider some specific cases of this combination process, subject always to the following very basic assumption:

The resultant of two or more harmonic vibrations will be taken to be simply the sum of the individual vibrations. In the present discussion we are treating this as a purely mathematical problem. Ultimately, however, it becomes a *physical* question: Is the displacement produced by two disturbances, acting together, equal to the straightforward superposition of the displacements as they would be observed to occur separately? The answer to this question may be yes or no, according to whether or not the displacement is strictly proportional to the force producing it. If simple addition holds good, the system is said to be *linear*, and most of our discussions will be confined to such systems. As we have just said, however, we are for the moment addressing ourselves to the purely mathematical problem of adding two (or more) displacements, each of which is a sinusoidal function of time; the physical applicability of the results is not involved at this point.

TWO SUPERPOSED VIBRATIONS OF EQUAL FREQUENCY

Suppose we have two SHM's described by the following equations:

$$x_1 = A_1 \cos(\omega t + \alpha_1)$$
$$x_2 = A_2 \cos(\omega t + \alpha_2)$$

Their combination is then as follows:

$$x = x_1 + x_2 = A_1 \cos(\omega t + \alpha_1) + A_2 \cos(\omega t + \alpha_2) \qquad (2\text{--}1)$$

It is possible to express this displacement as a single harmonic vibration:

$$x = A \cos(\omega t + \alpha)$$

The rotating-vector description of SHM provides a very nice way of obtaining this result in geometrical terms. In Fig. 2–1(a) let OP_1 be a rotating vector of length A_1, making the angle $(\omega t + \alpha_1)$ with the x axis at time t. Let OP_2 be a rotating vector of length A_2 at the angle $(\omega t + \alpha_2)$. The sum of these is then the vector OP as defined by the parallelogram law of vector addition. As OP_1 and OP_2 rotate at the same angular speed ω, we can think of the parallelogram OP_1PP_2 as a rigid figure that rotates bodily at this same speed. The vector OP can be obtained as the vector

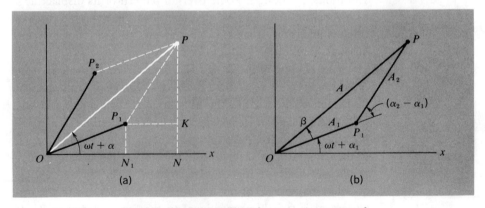

Fig. 2–1   (a) Superposition of two rotating vectors of the same period. (b) Vector triangle for constructing resultant rotating vector.

20   The superposition of periodic motions

sum of OP_1 and P_1P (the latter being equal to OP_2). Since $\angle N_1OP_1 = \omega t + \alpha_1$, and $\angle KP_1P = \omega t + \alpha_2$, the angle between OP_1 and P_1P is just $\alpha_2 - \alpha_1$. Hence we have

$$A^2 = A_1{}^2 + A_2{}^2 + 2A_1A_2 \cos(\alpha_2 - \alpha_1)$$

The vector OP makes an angle β [see Fig. 2–1(b)] with the vector OP_1, such that

$$A \sin\beta = A_2 \sin(\alpha_2 - \alpha_1)$$

and the phase constant α of the combined vibration is given by

$$\alpha = \alpha_1 + \beta$$

Use of the complex exponential formalism takes us, very directly, to these same results. The rotating vectors OP_1 and OP_2 are described by the following equations:

$$z_1 = A_1 e^{j(\omega t + \alpha_1)}$$
$$z_2 = A_2 e^{j(\omega t + \alpha_2)}$$

Hence the resultant is given by

$$z = z_1 + z_2 = A_1 e^{j(\omega t + \alpha_1)} + A_2 e^{j(\omega t + \alpha_2)}$$

Observe the advantage of using the exponential form, which allows us to take out the common factor $\exp j(\omega t + \alpha_1)$:

$$z = e^{j(\omega t + \alpha_1)}[A_1 + A_2 e^{j(\alpha_2 - \alpha_1)}] \tag{2-2}$$

Remembering that $e^{j\theta}$ is just an instruction to apply a positive rotation through the angle θ, we see that the combination of terms in square brackets specifies that a vector of length A_2 is to be added at an angle $(\alpha_2 - \alpha_1)$ to a vector of length A_1, and the initial factor $\exp[j(\omega t + \alpha_1)]$ tells us that this whole diagram is to be turned to the orientation shown in Fig. 2–1(b). If one did not take advantage of these geometrical techniques, the task of combining the two separate terms in Eq. (2–1) would be tiresome and much less informative.

In general the values of A and α for the resultant disturbance cannot be further simplified, but the special case in which the combining amplitudes are equal is worth noting. If we denote the *phase difference* $(\alpha_2 - \alpha_1)$ between the two vibrations as δ, then from the geometry of the vector triangle in Fig. 2–1(b) one can read off, more or less by inspection, the following results:

$$\beta = \frac{\delta}{2}$$

$$A = 2A_1 \cos\beta = 2A_1 \cos\left(\frac{\delta}{2}\right) \tag{2-3}$$

21   Two superposed vibrations of equal frequency

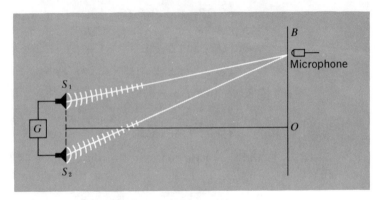

Fig. 2–2   *Array to detect phase difference as function of microphone position in the superposition of signals from two loudspeakers.*

A combination very much of this kind occurs if two identical loudspeakers are driven sinusoidally from the same signal generator and the sound vibrations are picked up by a microphone at a fairly distant point, as indicated in Fig. 2–2. If the microphone is moved along the line OB, the phase difference δ increases steadily from an initial value of zero at O. If the wavelength of the sound waves is much shorter than the separation of the speakers, the resultant amplitude A may be observed to fall to zero at several points between O and B, and rise to its maximum possible value of $2A_1$ at other points midway between the zeros. (We shall discuss such situations in more detail in Chapter 8.)

SUPERPOSED VIBRATIONS OF DIFFERENT FREQUENCY; BEATS

Let us now imagine that we have two vibrations of different

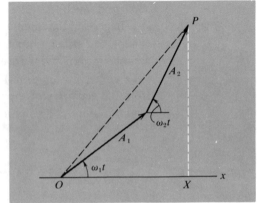

Fig. 2–3 *Superposition of rotating vectors of different periods.*

amplitudes A_1, A_2, and also of different angular frequencies ω_1, ω_2. Clearly, in contrast to the preceding example, the phase difference between the vibrations is continually changing. The specification of some initial nonzero phase difference is in general not of major significance in this case. To simplify the mathematics, let us suppose, therefore, that the individual vibrations have zero initial phase, and hence can be written as follows:

$$x_1 = A_1 \cos \omega_1 t$$
$$x_2 = A_2 \cos \omega_2 t$$

At some arbitrary instant the combined displacement will then be as shown (OX) in Fig. 2–3. Clearly the length OP of the combined vector must always lie somewhere between the sum and the difference of A_1 and A_2; the magnitude of the displacement

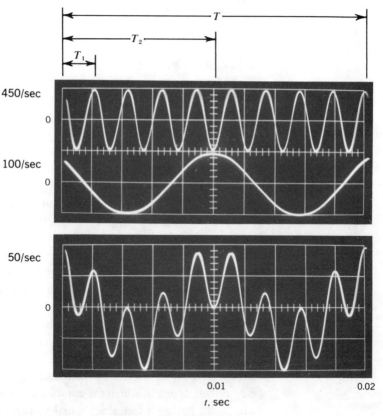

Fig. 2–4 *Superposition of two sinusoids with commensurable periods* ($T_1 = 1/450$ sec, $T_2 = 1/100$ sec.) *(Photo by Jon Rosenfeld, Education Research Center, M.I.T.).*

23   Superposed vibrations and beats

OX itself may be anywhere between zero and $A_1 + A_2$.

Unless there is some simple relation between ω_1 and ω_2, the resultant displacement will be a complicated function of time, perhaps even to the point of never repeating itself. The condition for any sort of true periodicity in the combined motion is that the periods of the component motions be commensurable—i.e., there exist two integers n_1 and n_2 such that

$$T = n_1 T_1 = n_2 T_2 \qquad (2\text{-}4)$$

The period of the combined motion is then the value of T as obtained above, using the smallest integral values of n_1 and n_2 for which the relation can be written.[1]

Even if the periods or frequencies are expressible as a ratio of two fairly small integers, the general appearance of the motion is not particularly simple. Figure 2–4 shows two component sinusoidal vibrations of 450 and 100 Hz, respectively. The repetition period is 0.02 sec, as may be inferred from the condition

$$T = \frac{n_1}{450} = \frac{n_2}{100}$$

which requires $n_1 = 9$, $n_2 = 2$, according to Eq. (2–4).

In those cases in which a vibration is built up of two commensurable periods, the appearance of the resultant may depend markedly on the relative initial phase of the combining vibrations. This effect is illustrated in Figs. 2–5(a) and (b), both of which make use, in the manner shown, of combining vibrations with given values of amplitude and frequency. Only the phase relationship differs in the two cases. Interestingly enough, if these were vibrations of the air falling upon the eardrum, the aural effects of the two combinations would be almost indistinguishable. It appears that the human ear is rather insensitive to phase in a mixture of harmonic vibrations; the amplitudes and frequencies dominate the situation, although significantly different aural effects may be produced if the different phase relationships lead to drastically different waveforms, as can happen if many frequencies, rather than just two, are combined with particular phase relationships.

If two SHM's are quite close in frequency, the combined disturbance exhibits what are called *beats*. This phenomenon can be described as one in which the combined vibration is basically a disturbance having a frequency equal to the average of the two

---

[1]If, for example, the ratio ω_1/ω_2 were an irrational (e.g., $\sqrt{2}$), there would exist no time, however long, after which the preceding pattern of displacement would be repeated.

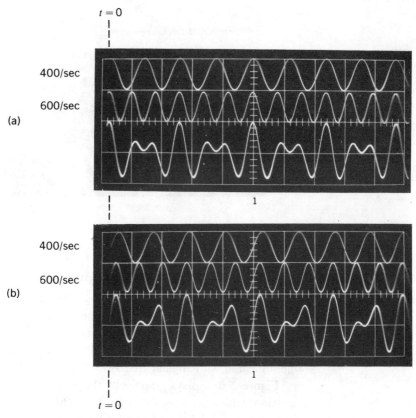

Fig. 2–5 (a) Superposition of two commensurable sinusoids, of frequencies 400 sec⁻¹ and 600 sec⁻¹, whose maxima coincide at t = 0. (b) Superposition of same sinusoids if their zeros coincide at t = 0. (Photos by Jon Rosenfeld, Education Research Center, M.I.T.)

combining frequencies, but with an amplitude that varies periodically with time—one cycle of this variation including many cycles of the basic vibration.

The beating effect is most easily analyzed if we consider the addition of two SHM's of equal amplitude:

$$x_1 = A \cos \omega_1 t$$
$$x_2 = A \cos \omega_2 t$$

Then by addition we get[1]

$$x = 2A \cos\left(\frac{\omega_1 - \omega_2}{2} t\right) \cos\left(\frac{\omega_1 + \omega_2}{2} t\right) \tag{2-5}$$

---

[1]You may wish to recall the following trigonometric results:

$$\cos(\theta + \varphi) = \cos\theta \cos\varphi - \sin\theta \sin\varphi$$
$$\cos(\theta - \varphi) = \cos\theta \cos\varphi + \sin\theta \sin\varphi$$

(continued)

25   Superposed vibrations and beats

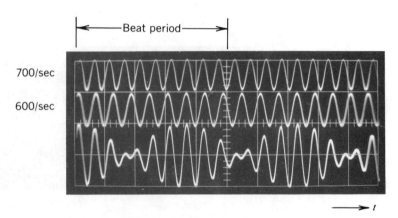

700/sec

600/sec

$\longrightarrow t$

Fig. 2–6 Superposition of sinusoids of nearly equal frequency ($600\ sec^{-1}$ and $700\ sec^{-1}$) to produce beats. (Photo by Jon Rosenfeld, Education Research Center, M.I.T.)

Clearly this addition, as a purely mathematical result, can be carried out for *any* values of ω_1 and ω_2. But its description as a beat phenomenon is physically meaningful only if $|\omega_1 - \omega_2| \ll \omega_1 + \omega_2$; i.e., if, over some substantial number of cycles, the vibration approximates to sinusoidal vibration with constant amplitude and with angular frequency $(\omega_1 + \omega_2)/2$.

Figure 2–6 displays graphically the result of combining two vibrations with a frequency ratio of $7:6$. This is about as large a ratio as one could have and still refer to the combination as a beat vibration. It may be seen that the combined displacement can be fitted within an envelope defined by the pair of equations

$$x = \pm 2A \cos\left(\frac{\omega_1 - \omega_2}{2}t\right) \tag{2–6}$$

because the rapidly oscillating factor in Eq. (2–5)—i.e., $\cos(\omega_1 + \omega_2)t/2$—always lies between the limits ± 1, and Eq. (2–6) describes a relatively slow amplitude-modulation of this oscillation. If one refers to Fig. 2–6, one sees that the time between successive zeros of the modulating disturbance is *one half-period* of the modulating factor as described by Eq. (2–6), i.e., a time equal to $2\pi/(|\omega_1 - \omega_2|)$. This has the consequence that

Therefore,
$$\cos(\theta + \varphi) + \cos(\theta - \varphi) = 2\cos\theta\cos\varphi$$
In this identity, let $\theta + \varphi = \alpha$, $\theta - \varphi = \beta$. Then
$$\cos\alpha + \cos\beta = 2\cos\left(\frac{\alpha + \beta}{2}\right)\cos\left(\frac{\alpha - \beta}{2}\right)$$
In the case under discussion, we put $\alpha = \omega_1 t$, $\beta = \omega_2 t$.

26   The superposition of periodic motions

the *beat frequency*—as observed aurally, e.g., from two tuning forks—is simply the difference of their individual frequencies and not half this frequency, as might be suggested by a first glance at Eq. (2–5). Thus, to take a specific case, if two tuning forks side by side are vibrating at 255 and 257 vibrations per second, their combined effect would be that of middle C (256 vibrations per second) passing through a maximum of loudness twice every second.

MANY SUPERPOSED VIBRATIONS OF THE SAME FREQUENCY[1]

The procedures that we have been describing can readily be extended to an arbitrarily large number of combining vibrations. The general case is of no great importance, but one situation, in particular, is of great interest and wide application. It is the case in which one has a superposition of a number of SHM's, all of the same frequency and amplitude, and with equal successive phase differences. This problem has special relevance to the analysis of multiple-source interference effects in optics and other wave processes.

The situation is represented in Fig. 2–7. We suppose that there are N combining vibrations, each of amplitude A_0 and differing in phase from the next one by an angle δ. Let the first

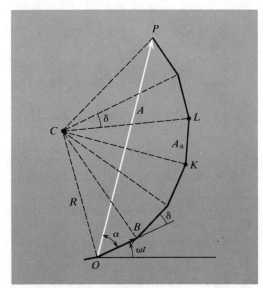

Fig. 2–7 *Superposition of several rotating vectors of same period and constant incremental phase differences.*

[1]This section may be omitted without loss of continuity.

of the component vibrations be described, for simplicity, by the equation

$$x = A_0 \cos \omega t$$

The resultant disturbance will be given by the equation

$$X = A \cos(\omega t + \alpha)$$

From the geometry of Fig. 2–7, we can see that the combining vectors form successive sides of an (incomplete) regular polygon. Any such polygon can be imagined to be inscribed in a circle, having some radius R and with its center at a point C. All the corners (as, for example, the points K and L) lie on the circle, and the angle subtended at C by any individual amplitude A_0 (e.g., KL) is equal to the angle δ between adjacent vectors. Hence the total angle OCP, subtended at C by the resultant vector A, is equal to $N\delta$. We can then write the following geometrical statements:

$$A = 2R \sin(N\,\delta/2)$$
$$A_0 = 2R \sin(\delta/2)$$

Therefore,

$$A = A_0 \frac{\sin(N\,\delta/2)}{\sin(\delta/2)} \qquad (2\text{--}7)$$

Also, for the phase angle α through which the resultant A is rotated relative to the first component vector, we have

$$\alpha = \angle COB - \angle COP$$

with

$$\angle COB = 90° - \frac{\delta}{2}$$

$$\angle COP = 90° - \left(\frac{N\delta}{2}\right)$$

Therefore,

$$\alpha = \frac{(N-1)\delta}{2} \qquad (2\text{--}8)$$

Hence the resultant vibration along the x axis is described by the following equation:

$$X = A_0 \frac{\sin(N\,\delta/2)}{\sin(\delta/2)} \cos\left[\omega t + \frac{(N-1)\delta}{2}\right] \qquad (2\text{--}9)$$

This equation is basic to the analysis of the behavior of a diffraction grating, which acts precisely as a device to obtain from a single beam of light a very large number of equal disturbances with equal phase differences.

COMBINATION OF TWO VIBRATIONS AT RIGHT ANGLES

Everything we have discussed so far has been concerned with harmonic motion along one physical dimension only, even though in analyzing it we have introduced the helpful concept of a vector rotating in a plane, such that the projection of the vector on a certain defined direction should represent the actual motion. We shall now discuss the *essentially different* problem of combining two real harmonic vibrations that take place along perpendicular directions, so that the resultant real motion is a true two-dimensional motion. This is a problem of considerable physical interest, and is appropriately discussed here because the analysis of it draws upon the same techniques that we have been using earlier in this chapter. The type of motion that we are about to discuss can be extended in a straightforward way to three-dimensional oscillations, such as one must in general suppose possible—as, for example, in the case of an atom elastically

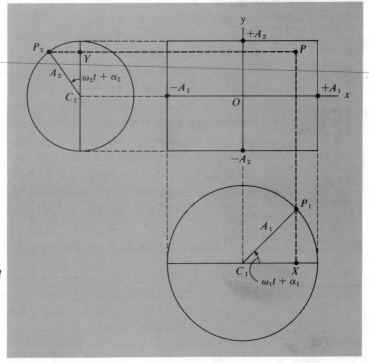

Fig. 2–8 Geometrical representation of the superposition of simple harmonic vibrations at right angles.

bound in the essentially three-dimensional structure of a crystal lattice.

We now suppose, therefore, that a point experiences the following displacements simultaneously:

$$x = A_1 \cos(\omega_1 t + \alpha_1)$$
$$y = A_2 \cos(\omega_2 t + \alpha_2)$$

(2-10)

We can construct this motion by means of a double application of the rotating-vector technique. The way of doing this is displayed in Fig. 2-8. We begin by drawing two circles, of radii A_1 and A_2, respectively. The first is used to define the x displacement $C_1 X$ of the point P_1. The second is used to define the y displacement $C_2 Y$ of the point P_2. The two displacements together describe the instantaneous position of the point P with respect to an origin O that lies at the center of a rectangle of sides $2A_1$ and $2A_2$.

One feature is immediately apparent. Whatever the relation between the frequencies and the phases of the two combining motions, the motion of the point P is always confined within the rectangle, and also the sides of this rectangle are tangential to the path at every point at which the path touches these boundary lines.[1] We cannot say much more than this without specifying something about the frequencies and phases, except for a general comment about what happens if ω_1 and ω_2 are not commensurable. In any such case, the position of P will never repeat itself, and the path, if continued for long enough, will, from a physical standpoint even if not from a strictly mathematical one, tend to fill the whole interior of the bounding rectangle.

The most interesting examples of these combined motions are those for which the frequencies are in some simple numerical ratio and the difference of the initial phases is some simple fraction of 2π. One then has a motion that forms a closed curve in two dimensions, with a period that is the lowest common multiple of the individual periods. The problem is best discussed in terms of specific examples, so let us look at a few.

PERPENDICULAR MOTIONS WITH EQUAL FREQUENCIES

By a suitable choice of what we call $t = 0$, we can write the combining vibrations in the following simple form:

[1]Except, perhaps, when the resultant motion goes into the corners of the rectangle, in which case the geometric conditions at the corners are not clearly defined.

$$x = A_1 \cos \omega t$$
$$y = A_2 \cos(\omega t + \delta)$$

where δ is thus the initial phase difference (and in this case the phase difference at all later times, too) between the motions. By specializing still further, to particular values of δ, we can quickly build up a qualitative picture of all possible motions for which the combining frequencies are equal:

a. $\delta = 0$. In this case,

$$x = A_1 \cos \omega t$$
$$y = A_2 \cos \omega t$$

Therefore,

$$y = \frac{A_2}{A_1} x$$

The motion is rectilinear, and takes place along a diagonal of the rectangle such that x and y always have the same sign, both positive or both negative. This represents what in optics is called a linearly polarized vibration.

b. $\delta = \pi/2$. We now have

$$x = A_1 \cos \omega t$$
$$y = A_2 \cos(\omega t + \pi/2) = -A_2 \sin \omega t$$

The shape of this path is readily obtained by making use of the fact that $\sin^2 \omega t + \cos^2 \omega t = 1$. This means that

$$\frac{x^2}{A_1{}^2} + \frac{y^2}{A_2{}^2} = 1$$

which is the equation of an ellipse whose principal axes lie along the x and y axes.

Notice, however, that the equations tell us more than this. We are dealing with kinematics, not geometry, and the ellipse is described in a definite direction. As t begins to increase from zero, x begins to decrease from its greatest positive value, and y immediately begins to go negative, starting from zero. This means that the elliptical path takes place in the *clockwise* direction.

c. $\delta = \pi$. We now have

$$x = A_1 \cos \omega t$$
$$y = A_2 \cos(\omega t + \pi) = -A_2 \cos \omega t$$

Therefore,

$$y = -\frac{A_2}{A_1} x$$

This motion is like case a, but is along the other diagonal of the rectangle.

d. $\pi = 3\pi/2$. This gives us

$$x = A_1 \cos \omega t$$

$$y = A_2 \cos \left(\omega t + \frac{3\pi}{2} \right) = +A_2 \sin \omega t$$

We have an ellipse of the same form as in case b, but the motion is now *counterclockwise*.

e. $\delta = \pi/4$. Note that we are jumping back here to the case of a phase difference between 0 and $\pi/2$, i.e., intermediate between cases a and b. It is a less obvious case than those just discussed, and lends itself to the graphical construction of Fig. 2–8. The application of the method to this particular case is shown in Fig. 2–9. The positions of the points P_1, P_2, on the two reference circles are shown at a number of instants separated by one eighth of a period (i.e., $\pi/4\omega$). The points are numbered in sequence, beginning with $t = 0$, when C_1P_1 (see Fig. 2–8) is parallel to the x axis and C_2P_2 is at the angle δ, i.e., 45°, measured counterclockwise from the positive y axis. The projections from

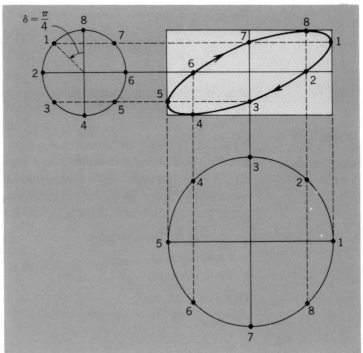

Fig. 2–9 *Superposition of simple harmonic vibrations at right angles with initial phase difference of $\pi/4$.*

32   The superposition of periodic motions

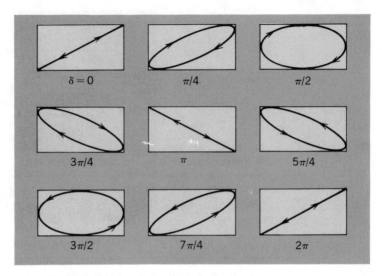

Fig. 2–10   *Superposition of two perpendicular simple harmonic motions of the same frequency for various initial phase differences.*

these corresponding positions of P_1 and P_2 then give us a set of intersections, as shown in Fig. 2–9, representing the instantaneous positions of the point P as it moves within the rectangle. The locus defined by these points is an ellipse, on inclined axes, described *clockwise*. The analytic equation of this ellipse can be found, if desired, by eliminating t from the defining equations for x and y:

$$x = A_1 \cos \omega t$$
$$y = A_2 \cos(\omega t + \pi/4)$$
$$= \frac{A_2}{\sqrt{2}} \cos \omega t - \frac{A_2}{\sqrt{2}} \sin \omega t$$

With the help of this last example, we can see how the pattern of this combined motion develops as we imagine the phase difference δ to increase from zero to 2π. Starting out from the linear diagonal motion at $\delta = 0$, the motion becomes a clockwise elliptical motion, opening up to a maximum width for $\delta = \pi/2$, and then closing down until at $\delta = \pi$ we have linear motion along the other diagonal. Beyond $\delta = \pi$ we pass through a similar sequence of elliptical motions (all of them now counterclockwise, however) until at $\delta = 2\pi$ we are back at a situation indistinguishable from $\delta = 0$. This sequence of motions is illustrated in Fig. 2–10.

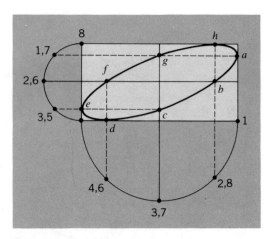

Fig. 2–11 Abbreviated construction for superposition of vibrations at right angles.

In all such problems the graphical method provides an excellent way of constructing the resultant motion. As in the example last discussed, the procedure is to mark on the reference circles a set of points corresponding to successive equal increments of time, and in particular to convenient submultiples of the period, such as eighths or twelfths or sixteenths. Once one is familiar with the process involved, one can make a more compact diagram by taking the bounding rectangle and simply constructing a semicircle on two adjacent sides. To illustrate this, let us take the case $\omega_1 = \omega_2$, $\delta = \pi/4$, once again. With the division of the reference circles into even submultiples of 2π, two different points on the circle project to give the same value of the displacement. Thus by drawing just a semicircle, one can convey as much information as with the full circle, but many of the points are used twice over, as indicated in Fig. 2–11. Once the points on the reference circles have been numbered according to the correct time sequence, the intersections that define the coordinate of the actual motion are obtained just as before. (To avoid confusion in this more condensed version of the diagram, we have used letters rather than numbers to identify these intersections— $a = 1$, $b = 2$, etc.)

Even if the instants chosen do *not* correspond to even submultiples of the total period (or even to *equal* submultiples) we can still indicate on the semicircle the correct sequence of points corresponding to one complete tour around the reference circle. It just involves imagining that the circle has been folded in half along its principal diameter—i.e., the diameter parallel to the component of the motion that this circle describes. But it economizes effort to pair off the points as we have done above.

PERPENDICULAR MOTIONS WITH DIFFERENT FREQUENCIES; LISSAJOUS FIGURES

It is a simple exercise, and a quite entertaining one, to extend the above analysis to motions with different frequencies. We give a few examples to illustrate the kind of results obtained.

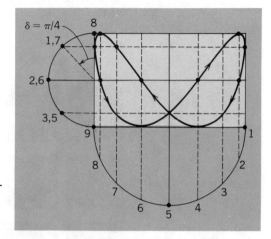

Fig. 2–12 Construction of a Lissajous figure.

In Fig. 2–12 we show the construction that one can make if $\omega_2 = 2\omega_1$ and $\delta = \pi/4$. We have chosen to divide the reference circle for the motion of frequency ω_2 into eight equal time intervals, i.e., into arcs subtending 45° each. During one complete cycle of ω_2, we go through only a half-cycle of ω_1, and the points on the reference circles are marked accordingly, taking account of the assumed initial phase difference of 45°. To obtain one complete period of the combined motion it is, of course, necessary to go through a complete cycle of ω_1; this requires that, after reaching the point marked "9," we retrace our steps along the lower semicircle and proceed for a second time through all the points corresponding to a complete tour around the ω_2 circle. In this way we end up with a closed path which crosses itself at one point and would be indefinitely repeated. Such a curve is known as a Lissajous figure, after J. A. Lissajous (1822–1880), who made an extensive study of such motions. If one introduces

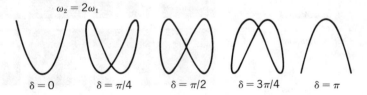

Fig. 2–13 Lissajous figures for $\omega_2 = 2\omega_1$ with various initial phase differences.

$\omega_2 = 2\omega_1$

$\delta = 0 \qquad \delta = \pi/4 \qquad \delta = \pi/2 \qquad \delta = 3\pi/4 \qquad \delta = \pi$

a slow decay of amplitude with time the patterns become still more exotic and have an esthetic appeal all their own. In Fig. 2–13 we show a set of such curves, all for $\omega_2 = 2\omega_1$, with initial phase differences of various sizes.

As one goes to more complicated frequency ratios, the resulting curves tend to become more bizarre, and Fig. 2–14 shows an assortment of examples. Such patterns are readily generated, with flexible control over amplitudes, frequencies, and phases, by applying different sinusoidal voltages to the x and y deflection

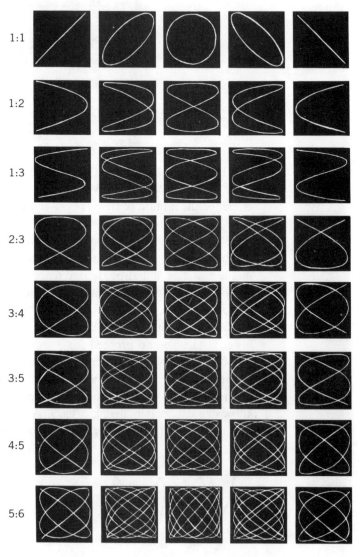

Fig. 2–14 Lissajous figures: various frequency ratios with various differences of phase. (After J. H. Poynting, J. J. Thomson, and W. S. Tucker, Sound, Griffin, London, 1949.)

The superposition of periodic motions

plates of a cathode-ray oscilloscope. Except in those cases where the Lissajous figure goes into the exact corners of the bounding rectangle, the ratio of the combining frequencies can be found by inspection; it is given by the ratio of the numbers of tangencies made by the figure with the adjacent sides of the rectangle. You should satisfy yourself of the theoretical justification of this result, and you can check its application to the various curves of Fig. 2–14.

COMPARISON OF PARALLEL AND PERPENDICULAR SUPERPOSITION

It is perhaps instructive to make a direct comparison of the superposition of two harmonic vibrations along the same line, and the superposition of the same vibrations in the orthogonal arrangement that leads to Lissajous figures. We have tried to display this relationship in Fig. 2–15, for the simple case of two vibrations of the same frequency and equal amplitudes. The figure shows two sinusoidal vibrations combined for various phase differences between zero and π. The lowest two curves of each group show the individual original displacements as y deflections on a double-beam oscilloscope with a linear time base. Above each pair of curves is the sinusoid resulting from the direct addition of these two y deflections. Finally, we show the Lissajous pattern obtained by switching off the time base of the oscilloscope and applying the two primary sinusoidal signals to the x and y plates.

If the two primary signals are given by $A \cos \omega t$ and $A \cos (\omega t + \delta)$, we have the following results:

Parallel Superposition

$$y_1 = A \cos \omega t$$
$$y_2 = A \cos (\omega t + \delta)$$
$$y = y_1 + y_2 = \left(2A \cos \frac{\delta}{2}\right) \cos \left(\omega t + \frac{\delta}{2}\right)$$

[Note the smooth decrease of amplitude in proportion to $\cos (\delta/2)$ as δ increases from zero to π.]

Perpendicular Superposition

$$x = A \cos \omega t$$
$$y = A \cos (\omega t + \delta)$$

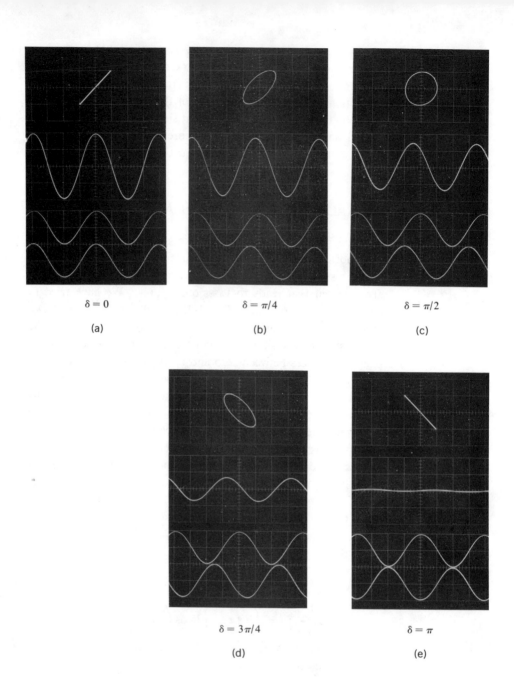

$\delta = 0$

(a)

$\delta = \pi/4$

(b)

$\delta = \pi/2$

(c)

$\delta = 3\pi/4$

(d)

$\delta = \pi$

(e)

Fig. 2–15   Comparison of the results of adding two harmonic vibrations (a) along the same line; (b) at right-angles to form Lissajous patterns.   (Photos by Jon Rosenfeld, Education Research Center, M.I.T.).

38   The superposition of periodic motions

Eliminating the explicit time dependence, we have

$$x^2 - 2xy \cos \delta + y^2 = A^2 \sin^2 \delta,$$

defining an elliptic curve which degenerates into a straight line for $\delta = 0$ or π, and into a circle for $\delta = \pi/2$, as shown in the photographs.

PROBLEMS

2-1 Express the following in the form $z = \text{Re}[Ae^{i(\omega t + \alpha)}]$
(a) $z = \sin \omega t + \cos \omega t$.
(b) $z = \cos(\omega t - \pi/3) - \cos \omega t$.
(c) $z = 2 \sin \omega t + 3 \cos \omega t$.
(d) $z = \sin \omega t - 2 \cos(\omega t - \pi/4) + \cos \omega t$.

2-2 A particle is simultaneously subjected to three simple harmonic motions, all of the same frequency and in the x direction. If the amplitudes are 0.25, 0.20, and 0.15 mm, respectively, and the phase difference between the first and second is 45°, and between the second and third is 30°, find the amplitude of the resultant displacement and its phase relative to the first (0.25-mm amplitude) component.

2-3 Two vibrations along the same line are described by the equations

$$y_1 = A \cos 10\pi t$$
$$y_2 = A \cos 12\pi t$$

Find the beat period, and draw a careful sketch of the resultant disturbance over one beat period.

2-4 Find the frequency of the combined motion of each of the following:
(a) $\sin(2\pi t - \sqrt{2}) + \cos(2\pi t)$.
(b) $\sin(12\pi t) + \cos(13\pi t - \pi/4)$.
(c) $\sin(3t) - \cos(\pi t)$.

2-5 Two vibrations at right angles to one another are described by the equations

$$x = 10 \cos(5\pi t)$$
$$y = 10 \cos(10\pi t + \pi/3)$$

Construct the Lissajous figure of the combined motion.

2-6 Construct the Lissajous figures for the following motions:
(a) $x = \cos 2\omega t$, $y = \sin 2\omega t$.
(b) $x = \cos 2\omega t$, $y = \cos(2\omega t - \pi/4)$.
(c) $x = \cos 2\omega t$, $y = \cos \omega t$.

It is very evident that the Rule or Law of Nature in every springing body is, that the force or power thereof to restore it self to its natural position is always proportionate to the Distance or space it is removed therefrom, whether it be by rarefaction, or separation of its parts the one from the other, or by a Condensation, or crowding of those parts nearer together. Nor is it observable in these bodies only, but in all other springy bodies whatsoever, whether Metal, Wood, Stones, baked Earths, Hair, Horns, Silk, Bones, Sinews, Glass and the like. Respect being had to the particular figures of the bodies bended, and to the advantageous or disadvantageous ways of bending them.

ROBERT HOOKE, *De Potentia Restitutiva* (1678)

3

The free vibrations of physical systems

IN MAKING THE STATEMENT quoted opposite about the elastic properties of objects, Robert Hooke rather overstated the case. The restoring forces in any actual physical system are only *approximately* linear functions of displacement, as we noted near the beginning of Chapter 1. Nevertheless, it is remarkable that a vast variety of deformations of physical systems, involving stretching, compressing, bending, or twisting (or combinations of all of these) result in restoring forces proportional to displacement and hence lead to simple harmonic vibration (or a superposition of harmonic vibrations). In this chapter we shall consider a number of examples of such motions, with particular emphasis on the way in which we can relate the kinematic features of the motion to properties that can often be found by purely static measurement. We shall begin with a closer look at the system that forms a prototype for so many oscillatory problems— a mass undergoing one-dimensional oscillations under the type of restoring force postulated by Hooke. Much of the discussion in the next section will probably be familiar ground, but it is important to be quite certain of it before proceeding further.

THE BASIC MASS–SPRING PROBLEM

In our first reference to this type of system in Chapter 1, we characterized it as consisting of a single object of mass m acted

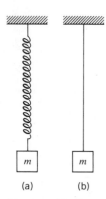

(a)   (b)

Fig. 3–1   (a) Mass–
spring system.
(b) Mass–wire system.

on by a spring [Fig. 3–1(a)] or some equivalent device, e.g., a thin wire [Fig. 3–1(b)], that supplies a restoring force equal to some constant k times the displacement from equilibrium. This identifies, in terms of a system of a particularly simple kind, the two features that are essential to the establishment of oscillatory motions:

1. An inertial component, capable of carrying kinetic energy.
2. An elastic component, capable of storing elastic potential energy.

By assuming that Hooke's law holds we obtain a potential energy proportional to the square of the displacement of the body from equilibrium. By assuming that the whole inertia of the system is localized in the mass at the end of the spring, we obtain a kinetic energy equal to just $mv^2/2$, where v is the speed of the attached object. It should be noted that both of these assumptions are specializations of the general conditions 1 and 2, and there will be many instances of oscillatory systems to which these special conditions do not apply. If, however, a system *can* be regarded as being effectively a concentrated mass at the end of a linear spring ("linear" referring to its elastic property rather than to its geometry), then we can write its equation of motion in either of two ways:

1. By Newton's law ($F = ma$),

$-kx = ma$

2. By conservation of total mechanical energy (E),

$\frac{1}{2}mv^2 + \frac{1}{2}kx^2 = E$

The second is, of course, the result of integrating the first with respect to the displacement x, but both of them are *differential equations* for the motion of the system. It is important to be able to recognize such differential equations wherever they emerge from the analysis of a physical system. In explicit differential form, they may be written as follows:

$$m\frac{d^2x}{dt^2} + kx = 0 \qquad\qquad (3\text{–}1)$$

$$\frac{1}{2}m\left(\frac{dx}{dt}\right)^2 + \frac{1}{2}kx^2 = E \qquad\qquad (3\text{–}2)$$

Whenever one sees an equation analogous to either of the above, one can conclude that the displacement x as a function of time is of the form

42   The free vibrations of physical systems

$$x = A \cos(\omega t + \alpha) \tag{3-3}$$

where ω^2 is the ratio (k/m) of the spring constant k to the inertia constant m. This will hold good, given Eq. (3–1) or (3–2), even if the system itself is not a single object on an effectively massless spring.

In Eq. (3–3) it is to be noted that the constant ω is defined for all circumstances by the given values of m and k. The equation contains two other constants—the amplitude A and the initial phase α—which between them provide a complete specification of the state of motion of the system at $t = 0$ (or other designated time) in any particular case. The initial statement of Newton's law in Eq. (3–1) contains *no* adjustable constants. Equation (3–2), often referred to as the "first integral" of Eq. (3–1), is mathematically intermediate between Eqs. (3–1) and (3–3) and contains *one* adjustable constant (the total energy E, which is equal to $kA^2/2$). The introduction of one more constant at each stage of integration of the original differential equation (Newton's law) is always necessary, even though in a particular case the constant may turn out to be zero. One can think of this as the reverse of the process whereby, in any differentiation, a constant term will disappear from sight.

SOLVING THE HARMONIC OSCILLATOR EQUATION USING COMPLEX EXPONENTIALS

As a pattern for future calculations, let us take the basic differential equation, Eq. (3–1), and develop the familiar solution as given in Eq. (3–3), making use of the complex exponential in the process. Since it is not k and m individually, but only the ratio k/m, that enters in any essential way, we begin by rewriting Eq. (3–1) in the following more compact form:

$$\frac{d^2x}{dt^2} + \omega^2 x = 0 \tag{3-4}$$

This states that x and its second time derivative are linearly combined to give zero, or equivalently that d^2x/dt^2 is a multiple of x itself. The exponential function is known to have this latter property; let us therefore put

$$x = Ce^{pt} \tag{3-5}$$

where (to have things dimensionally correct) we have introduced a coefficient C of the dimension of distance, and a coefficient p such that pt is dimensionless—i.e., p has the dimension of $(\text{time})^{-1}$.

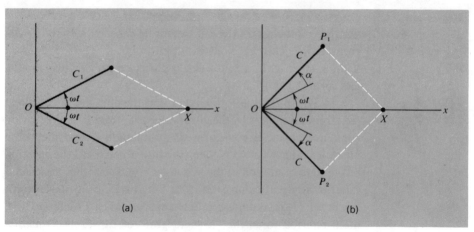

Fig. 3-2 (a) Superposition of complex solutions of
Eq. (3-4) with $C_1 = C_2$. (b) Superposition of complex
solutions of Eq. (3-4) for nonzero initial phase angle.

Then by substitution in Eq. (3-4) we have

$$p^2 C e^{pt} + \omega^2 C e^{pt} = 0$$

which can be satisfied for any t, and for any value of C, provided
that

$$p^2 + \omega^2 = 0$$

Therefore,

$$p^2 = -\omega^2$$
$$p = \pm j\omega \tag{3-6}$$

Each of these values of p will satisfy the original equation. Having
no reason to discard either, we accept both, each with its own
value of C. Thus Eq. (3-5) becomes

$$x = C_1 e^{j\omega t} + C_2 e^{-j\omega t} \tag{3-7}$$

Let us interpret Eq. (3-7) in terms of the rotating-vector
description of SHM. The first term on the right corresponds to
a vector C_1 rotating counterclockwise at angular speed ω, the
second term to a vector C_2 rotating clockwise at the same speed.
These combine to give a harmonic oscillation along the x axis,
as shown in Fig. 3-2(a), if the lengths of C_1 and C_2 are equal.
But C_1 and C_2, as they appear in Eq. (3-7), do not have to be
real. We can satisfy Eq. (3-7) just as well if C_1 is rotated through
some angle α with respect to the direction defined by t, provided
that C_2 is rotated through $-\alpha$ with respect to $-\omega t$, again making

44   The free vibrations of physical systems

the vectors of equal length, as shown in Fig. 3–2(b).[1] This less restrictive condition then leads to the customary result:

$$x = Ce^{i(\omega t+\alpha)} + Ce^{-i(\omega t+\alpha)}$$
$$= 2C\cos(\omega t + \alpha)$$
$$\equiv A\cos(\omega t + \alpha)$$

The quantities C_1 and C_2 in Eq. (3–7), or A and α in the above equation, represent equally well the two constants of integration that must be introduced in the process of going from the second-order differential equation (3.4)[2] to the final solution that expresses x itself as a function of t.

The above analysis reveals incidentally that a rectilinear harmonic motion can be produced by the superposition of two equal and opposite real circular motions—which is a kind of converse to the production of a circular Lissajous figure from two equal and perpendicular linear oscillations. (Both of these results have important applications in the description of polarized light.) Having arrived at the final equation, we see that x can be described as the real part of a rotating vector corresponding just to the first term alone in Eq. (3.7).[3] Thus in many future calculations we shall assume solutions simply of the following type:

$$x = \text{real part of } z \quad \text{where} \quad z = Ae^{i(\omega t+\alpha)} \tag{3–8}$$

This extended rediscussion of the simple harmonic oscillator, although it deals only with very familiar results, may help to provide some further insight into the workings of the rotating complex vector description of SHM, and into the justification of this approach.

ELASTICITY AND YOUNG'S MODULUS

Let us turn now to the properties of matter that control the frequency of a mass–spring type of system. If we consider an actual coiled spring the problem is a complicated one. The attachment of a load to such a spring, as shown in Fig. 3–3, gives rise to two different effects, neither of which is a simple stretching process. If we imagine a weight W suspended from a point on the vertical

---

[1]No other relationship leads to oscillation along the x axis alone. Try to satisfy yourself on this point.

[2]The *order* of a differential equation is defined by the highest derivative appearing in it.

[3]Or the second term alone, if preferred.

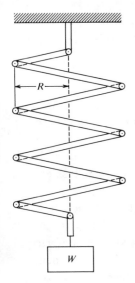

Fig. 3–3 Coiled
spring with suspended
mass.

axis of the coil, its effect is to produce a torque WR about any point on the approximately horizontal axis of the wire composing the spring. One effect of this—the chief effect in most springs— is to twist the wire about on its own axis, and the descent of the weight is primarily a consequence of this twisting process. But there is a second effect: The coils of the spring will tighten or loosen a little, so that the spring as a whole twists about the vertical axis. This process involves a bending of the coils—i.e., a change in their curvature.[1] The final result is, to be sure, expressible as a proportionality (the spring constant k) between the applied load and the distance through which the load moves, but in relating springiness to basic physical properties we shall do well to turn aside from the familiar coiled spring to more straight-forward problems.

The simple stretching of a rod or wire provides the most easily discussed situation of all. The behavior of such a system under conditions of static equilibrium can be described as follows:

1. For a given material made up into rods or wires of a given cross-sectional area, the extension Δl under a given force is proportional to the original length l_0. The dimensionless ratio $\Delta l/l_0$ is called the *strain*. This result can also be expressed by saying that in a *static* experiment with a given rod, the displacements of various points along it are proportional to their distances from the fixed end, as shown in Fig. 3–4(a), because in such a static situation the force ΔP applied at one end gives rise to a tension of magnitude ΔP along the whole length of the rod.

2. It is also found that, for rods of a given material, but of different cross-sectional areas, the same strain ($\Delta l/l_0$) is caused by applying forces proportional to the cross-sectional areas, as in Fig. 3–4(b). The ratio $\Delta P/A$ is called the *stress* and has the dimensions of force per unit area, or pressure.

3. Provided that the strain is very small—less than about 0.1% of the normal length l_0, the relation between stress and strain is linear, in accordance with Hooke's law. In this case we can write

$$\frac{\text{stress}}{\text{strain}} = \text{constant}$$

The value of this constant for any given material is called Young's modulus of elasticity (after the same Thomas Young who made scientific history in 1801 with his optical interference experi-

[1]Whether a spring will tighten or loosen depends on the material of which it is made.

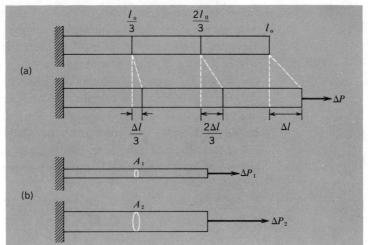

Fig. 3–4 (a) Uniform longitudinal extension of rod under static conditions. (b) Rods of different cross sections A_1 and A_2 under tensions ΔP_1 and ΔP_2.

ments[1]). It is usually given the symbol Y. If we denote by dF the force exerted *by* a stretched wire or rod *on* another object, we can thus put

$$\frac{dF/A}{dl/l_0} = -Y$$

i.e.,

$$dF = -\frac{AY}{l_0} dl \tag{3-9}$$

If we choose to denote the extension by x and the force by F, we can alternatively write this result as follows:

$$F = -\frac{AY}{l_0} x \tag{3-10}$$

which then corresponds to the usual statement of the restoring force due to a stretched springlike object and identifies the spring constant k as AY/l_0 in this case. Table 3–1 lists the approximate

TABLE 3–1:   TENSILE PROPERTIES OF MATERIALS

Material	Young's Modulus, N/m^2	Ultimate strength, N/m^2
Aluminum	6×10^{10}	2×10^8
Brass	9×10^{10}	4×10^8
Copper	12×10^{10}	5×10^8
Glass	6×10^{10}	10×10^8
Steel	20×10^{10}	11×10^8

[1]He also made important contributions to the first decipherment of Egyptian hieroglyphics on the famous Rosetta Stone.

47   Elasticity and Young's modulus

values of Young's modulus for some familiar solid materials. Also shown are approximate values of the ultimate strength, expressed as the stress at which the material is liable to fracture. Notice that the Young's modulus represents a stress corresponding to 100% elongation, a condition that is never approached in the actual stretching of a sample. Failure occurs at stresses two or three orders of magnitude less than this, i.e., at strain values of between 0.1 and 1%. There is no possibility of obtaining, by direct stretching of a wire or rod, the kind of large fractional change of length that one can achieve so readily with a coiled spring.

If a body of mass m is hung on the end of a wire, the period of oscillations of very small amplitude is given by

$$T = 2\pi\sqrt{\frac{ml_0}{AY}} \tag{3-11}$$

as one can see from the force law, Eq. (3–10). For example, consider a mass of 1 kg hung on a steel wire of length 1 m and diameter 1 mm. We have

$$A = \frac{\pi d^2}{4} \approx 0.8 \times 10^{-6}\,\mathrm{m}^2$$

Therefore,

$$k = \frac{AY}{l_0} \approx 1.6 \times 10^5\,\mathrm{N/m}$$

Therefore,

$$T \approx \frac{2\pi}{400} \approx 1.6 \times 10^{-2}\,\mathrm{sec}$$

or

$$\nu = \frac{1}{T} \approx 60\,\mathrm{Hz}$$

One sees that this wire acts as a very hard spring, and the oscillations, besides being of quite high frequency, must also be of very small amplitude—only a small fraction of a millimeter in a 1-m wire—if the strength limit of the material is not to be exceeded.

The result expressed in Eq. (3–11) can be rewritten in a physically more vivid way if we introduce the *increase of length*, h, that occurs in *static* equilibrium when the body of mass m is first hung onto the wire. We have, by Eq. (3–10),

$$mg = \frac{AY}{l_0}h$$

Therefore,

$$\frac{ml_0}{AY} = \frac{h}{g}$$

Hence, from Eq. (3–11), we have

$$T = 2\pi \sqrt{\frac{h}{g}} \tag{3–12}$$

Thus the period is the same as that of a simple pendulum of length h. This makes a very straightforward way of computing the period on the basis of a single measurement of static extension, without any need for detailed knowledge of the characteristics of the wire or the magnitude of the attached mass.

The macroscopic elastic property described by Young's modulus must, of course, be analyzable in terms of the microscopic interactions between atoms in the material. Clearly, if the over-all length of a wire increases by 1%, this means that the individual interatomic spacings along that direction also increase by 1%. Thus one can, in principle, relate the elastic modulus to atomic properties as described by the potential-energy curve of the interatomic forces. We shall not, however, pursue that line of discussion here, because our immediate concern is with the macroscopic description. Instead, we shall proceed to the discussion of some other examples of simple harmonic motion.

FLOATING OBJECTS

If a floating object is slightly depressed or raised from its normal position of equilibrium, there is called into play a restoring force equal to the increase or decrease in the weight of liquid displaced by the object, and periodic motion ensues. The situation becomes especially simple if the floating body has a constant cross-sectional area over the part that intersects the liquid surface. A hydrometer (Fig. 3–5), as used to measure the specific gravity of battery acid or antifreeze, is a nice practical example of this.

Let the mass of the hydrometer be m, and let the liquid density be ρ. Denote the area of cross section as A. Then if the hydrometer is at distance y above its normal floating level, the volume of liquid displaced is equal to Ay and the equation of motion (Newton's law) becomes

$$m\frac{d^2y}{dt^2} = -g\rho Ay$$

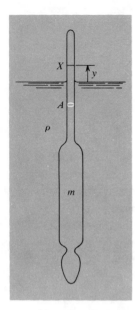

giving

$$\omega = \sqrt{\frac{g\rho A}{m}} \quad \text{and} \quad T = 2\pi \sqrt{\frac{m}{g\rho A}} \qquad (3\text{-}13)$$

For example, a common type of battery hydrometer has $m \approx 10$ g, $A \approx 0.25$ cm^2. Suppose that it is placed in battery acid of specific gravity 1.2. Then (using MKS units) we have

$$m \approx 10^{-2} \text{ kg}$$
$$A \approx 2.5 \times 10^{-5} \text{ m}^2$$
$$g \approx 10 \text{ m/sec}^2$$
$$\rho \approx 1.2 \times 10^3 \text{ kg/m}^3$$

which gives

$$T \approx 1 \text{ sec}$$

On a much larger scale, one can consider such motion occurring with a ship. To some approximation the sides of a big ship are almost vertical, and its bottom more or less flat, as in Fig. 3–6. In this case we can very conveniently express the mass of the ship in terms of its draft, h:

$$m = \rho A h$$

where ρ is the density of water and A is the horizontal cross-sectional area of the ship at the waterline. Substituting this in Eq. (3–12) we find

$$T = 2\pi \sqrt{\frac{h}{g}} \qquad (3\text{-}14)$$

which is thus exactly like the simple pendulum equation that also could be used for the vertical oscillations of a mass hung on a wire [Eq. (3–12)]. If, for example, the draft of the ship is 10 m, the period of such vertical oscillations would be about 6 sec.

Fig. 3-5 Simple hydrometer, capable of vertical oscillations when displaced from normal floating position.

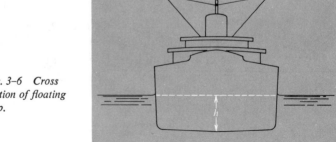

Fig. 3–6 Cross section of floating ship.

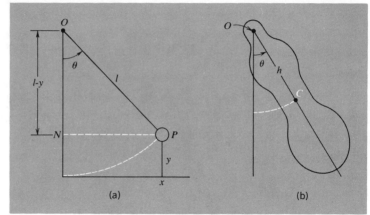

Fig. 3–7   (a) Simple pendulum. (b) Suspended mass of arbitrary shape on a horizontal axis (rigid pendulum.)

Such a motion would not, however, be an important component of the ship's total pattern of oscillation. Rolling and pitching, which do not involve any important rise or fall of the position of the center of mass relative to the water surface, are more readily excited by the action of the waves.

PENDULUMS

The so-called "simple pendulum," as shown in Fig. 3–7(a), represents a familiar oscillatory system that is nevertheless a good deal more complicated than the one-dimensional oscillators that we have considered so far (although it must be admitted that, in discussing the vertical oscillations of floating objects, we have conveniently overlooked the tricky question of the motion of the displaced liquid).

The problem of the pendulum is essentially two-dimensional, even though the actual displacement is completely specified by a single angle θ. Although the displacements are predominantly horizontal, the motion depends in an essential way on the fact that there is a rise and fall of the center of mass, with associated changes of gravitational potential energy. The pendulum is, in fact, well suited to an analysis beginning with a statement of energy conservation, and because the final result is almost certainly familiar it provides a good example of this energy method, which is of very great value in the analysis of more complicated systems.

Referring now to Fig. 3–7(a), if the angle θ is small we have $y \ll x$ and hence, from the geometry of the figure,

$$y \approx \frac{x^2}{2l}$$

where l is the length of the string.[1] The statement of conservation of energy is

$$\tfrac{1}{2}mv^2 + mgy = E \quad \text{where} \quad v^2 = \left(\frac{dx}{dt}\right)^2 + \left(\frac{dy}{dt}\right)^2$$

Given the approximations already introduced, it is thus very nearly correct to put

$$\frac{1}{2}\, m \left(\frac{dx}{dt}\right)^2 + \frac{1}{2}\frac{mg}{l}x^2 = E$$

which we recognize, according to Eq. (3–2), as defining simple harmonic motion with $\omega = \sqrt{g/l}$.

By way of preparation for more complicated pendulums, note the alternative statement of the problem in terms of the angular displacement θ. Using this, we have

$$v = l\left(\frac{d\theta}{dt}\right) \quad \text{(exactly)}$$

$$y = l(1 - \cos\theta) \approx \tfrac{1}{2}l\theta^2$$

so that our approximate statement of energy conservation is now

$$\tfrac{1}{2}ml^2 \left(\frac{d\theta}{dt}\right)^2 + \tfrac{1}{2}mgl\theta^2 = E$$

Consider now an arbitrary object that is free to swing in a vertical plane. Let its center of mass C be a distance h from the point of suspension, as shown in Fig. 3–7(b). Then the gain of potential energy for an angular deflection θ is $mgh\theta^2/2$. The kinetic energy is the energy of rotation of the body as a whole `bout O. Since every point in the body has angular speed $d\theta/dt$, kinetic energy can be written $I(d\theta/dt)^2/2$, where I is the moment of inertia about the horizontal axis through O. Hence we have

$$\tfrac{1}{2}I\left(\frac{d\theta}{dt}\right)^2 + \tfrac{1}{2}mgh\theta^2 = E$$

It is in many instances convenient to introduce the moment of inertia about a parallel axis through the center of mass. If this is written as mk^2, where k is the "radius of gyration" of the body,

---

[1] In the triangle ONP, we have (by Pythagoras's theorem) $l^2 = (l - y)^2 + x^2$. Hence $x^2 = 2ly - y^2 \approx 2ly$.

then the kinetic energy of rotation *with respect to the center of mass* is $mk^2(d\theta/dt)^2/2$, to which must be added the kinetic energy associated with the instantaneous linear speed $h(d\theta/dt)$ of the center of mass itself. Thus the energy-conservation equation may also be written as follows:

$$\tfrac{1}{2}mk^2\left(\frac{d\theta}{dt}\right)^2 + \tfrac{1}{2}m\left(h\frac{d\theta}{dt}\right)^2 + \tfrac{1}{2}mgh\theta^2 = E$$

from which we have

$$\omega^2 = \frac{gh}{h^2 + k^2}$$
$$T = 2\pi\left(\frac{h^2 + k^2}{gh}\right)^{1/2} \tag{3–15}$$

WATER IN A U-TUBE

If a liquid is contained in a U-tube arrangement of constant cross section with vertical arms, as shown in Fig. 3–8, we have a system that resembles the pendulum in that, although the motion is two-dimensional, it can be completely described in terms of the single vertical displacement y of the liquid surface from equilibrium.[1] Suppose that the total length of liquid column is l and its cross section is A. Then, if ρ is the liquid density, the total mass m of liquid is ρAl. We shall assume that every part of the liquid moves with the same speed, dy/dt. The increase of gravitational potential energy in the situation shown in Fig. 3–8 corresponds to taking a column of liquid of length y from the left-hand tube,

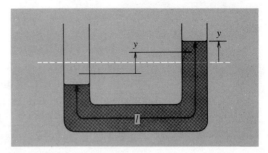

Fig. 3–8 Oscillating liquid column in a U-tube.

[1]The side arms need not, in fact, be vertical, as long as they are straight; the cross sections need not be equal, as long as they are constant; and the connecting tubing may be of different cross section again, provided the appropriate geometrical scaling factors are used to express the displacement and speed of any part of the liquid in terms of those in either of the side arms.

raising it through the distance y, and placing it on the top of the right-hand column. Thus we can put

$$U = g\rho A y^2$$

The conservation of mechanical energy thus gives the following equation:

$$\tfrac{1}{2}\rho A l \left(\frac{dy}{dt}\right)^2 + g\rho A y^2 = E$$

Hence

$$\omega^2 = \frac{2g}{l}$$

$$T = 2\pi \sqrt{\frac{l}{2g}} = \pi \sqrt{\frac{2l}{g}} \qquad\qquad (3\text{--}16)$$

Note the similarity to the simple pendulum equation, but also the subtle difference—that a liquid column in these circumstances has the same period as a simple pendulum of length $l/2$.

TORSIONAL OSCILLATIONS

The development of a restoring *torque*, and the existence of a stored potential energy in a twisted object, are familiar mechanical facts. If the torque M is proportional to the angular displacement between two ends of an object, we can put

$$M = -c\theta \qquad\qquad (3\text{--}17)$$

where c is the torsion constant of the system. The stored potential energy is thus given by

$$U = -\int M\,d\theta = \tfrac{1}{2}c\theta^2$$

If the angular deflection θ is given to a body of moment of inertia I attached to one end of the twisted system (and if the inertia of the twisted system itself is negligible), we then have an energy-conservation statement in the form

$$\tfrac{1}{2}I\left(\frac{d\theta}{dt}\right)^2 + \tfrac{1}{2}c\theta^2 = E$$

and hence

$$\omega^2 = \frac{c}{I}$$

$$T = 2\pi\sqrt{\frac{I}{c}} \qquad\qquad (3\text{--}18)$$

54 The free vibrations of physical systems

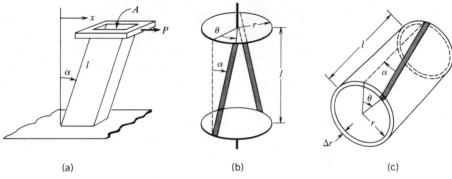

(a)   (b)   (c)

Fig. 3–9 (a) Shear deformation of rectangular block.
(b) Torque on rectangular strips during shear deforma-
tion. (c) A twisted tube can be thought of as a large
collection of strips such as those shown in (b).

The relation of the torsion constant to the basic elastic
properties of the twisted material is less direct than the relation
of a spring constant to Young's modulus for a stretched wire or
rod. The essential process is called a *shear* deformation of the
material. Suppose that a rectangular block of material is firmly
glued at its base to a table, and that its top face is glued to a flat
board [Fig. 3–9(a)]. Then a horizontal force P applied to the
board, in a direction parallel to two of the top edges of the block,
causes a deformation as shown.[1] Two of the side faces are
changed from rectangles into parallelograms. Thus the deformation can be characterized by the *angle of shear, α.* In terms of the
actual transverse displacement x of the top end of a block of
height l, we have (approximately)

$$\alpha = \frac{x}{l}$$

It is found that the angle of shear is proportional to the ratio of
the applied transverse force to the area A of the top surface of
the block. The proportionality of the *shear stress* P/A to the
angle of shear x/l is expressed by a *shear modulus,* or *modulus of
rigidity,* usually denoted by n. If we denote by $F (= -P)$ the
force exerted *by* the sheared material on the board, we can thus put

$$n = \frac{-F/A}{\alpha}$$

or

$$dF = -nA\,d\alpha = \frac{-nA}{l}\,dx \tag{3–19}$$

[1]The table, to maintain equilibrium, must supply a horizontal force $-P$ and
also a counterclockwise torque of magnitude lP, so that the block is not
subjected to any resultant translatory force or any resultant torque.

55 Torsional oscillations

These relations are thus of just the same type as we had for longitudinal deformations—Eqs. (3–9) and (3–10)—and the rigidity modulus n has the same physical dimensions as Young's modulus. For most materials these two moduli are of the same order of magnitude, although n is usually significantly less than Y. Table 3–2 shows values of both for the same selection of materials as in Table 3–1. Also shown is a third modulus—the so-called *bulk modulus*, K, which describes the resistance of a material to changes of volume.

TABLE 3–2: VALUES OF ELASTIC MODULI

Material	$Y, N/m^2$	$n, N/m^2$	$K, N/m^2$
Aluminum	6×10^{10}	3×10^{10}	7×10^{10}
Brass	9×10^{10}	3.5×10^{10}	6×10^{10}
Copper	12×10^{10}	4.5×10^{10}	13×10^{10}
Glass	6×10^{10}	2.5×10^{10}	4×10^{10}
Steel	20×10^{10}	8×10^{10}	16×10^{10}

To introduce the calculation of restoring torques from shearing processes, consider the situation shown in Fig. 3–9(b). Two disks of radius r on spindles are connected by a pair of rectangular strips of material. When one spindle is twisted through a small angle θ, the end of each strip is moved transversely through a distance $r\theta$. Thus the angle of shear is given by

$$\alpha = \frac{r\theta}{l}$$

If each strip has a cross-sectional area A, it provides a restoring force, tangential to the disk, given by

$$F = -nA\frac{r\theta}{l}$$

This then means a *torque*, of magnitude rF, exerted about the axis of twist by each strip.

Suppose now that one has a thin-walled tube, of mean radius r and wall thickness Δr, as shown in Fig. 3–9(c). This can be thought of as a whole collection of thin strips parallel to the axis of the cylinder, all contributing restoring torques about this axis. Thus the torque ΔM provided by the tube when its ends are given a relative twist θ is given by

$$\Delta M = -\frac{nAr^2\theta}{l}$$

where

$$A = 2\pi r \, \Delta r$$

Hence

$$\Delta M = -\frac{2\pi n r^3 \, \Delta r}{l} \theta$$

Finally if, as is most often the case, the twisted object is a solid cylindrical rod, wire, or fiber, the total torque is obtained by summing or integrating the above result. Hence we have

$$M = -\frac{\pi n r^4}{2l} \theta \qquad \text{(solid cylinder)} \qquad (3\text{–}20)$$

"THE SPRING OF AIR"

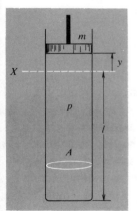

Fig 3–10   Piston in vertical air column.

One of our important topics in this book will be the analysis of the vibrations of air columns and the production of musical sounds. A useful foundation for this will be to consider a confined column of gas as being very much like a spring. Robert Boyle thought of the elasticity of a gas in just such terms, and the heading of this section is from the title of the book that he wrote about such matters.[1] (The word "spring," as Boyle used it, actually means the *quality* of springiness.)

To tie the discussion as closely as possible to our earlier analysis of the mass–spring system, suppose that we have a cylindrical tube, closed at one end, with a well-fitting but freely moving piston of mass m, as shown in Fig. 3–10. The entrapped column of air acts like a quite strong spring, very resistant to sudden pull or push; the effect is clearly demonstrated if one closes the exit hole of a bicycle pump with one finger and tries to move the plunger of the pump.

The piston has a certain equilibrium position, which will vary according to whether the tube is horizontal or vertical. If the tube is vertical, as shown in the figure, the pressure p of gas in the tube must be sufficiently above atmospheric to support the weight of the piston—just like the initial extension of a spring. But now, if the piston is moved a distance y, lengthening the air column, the internal pressure drops and the result is to provide a

[1]Robert Boyle, *New Experiments Physico-Mechanical Touching* [i.e., concerning] *the Spring of Air and Its Effects, Made for the Most Part in a New Pneumatical Engine*, Oxford, 1660.

restoring force on m. We can, in fact, write an equation of the form

$$F = A \Delta p$$

where Δp is the change of pressure.

How big is the pressure change? One's first thought might well be to calculate it from Boyle's law:

$$pV = \text{const.}$$

which would give us

$$p \Delta V + V \Delta p = 0 \qquad (3\text{--}21)$$

Now

$$\Delta V = Ay$$
$$V = Al$$

so that we should get

$$\Delta p = -\frac{py}{l}$$

and hence

$$F = -\frac{Ap}{l} y \qquad (3\text{--}22)$$

Compare this with Eq. (3–10) for the stretching or compressing of a solid rod. We see that in Eq. (3–22) the pressure p plays a role exactly analogous to an elasticity modulus. Indeed, given the assumption that Boyle's law applies, it *is* the elastic modulus of the air. It is not the Young's modulus, however, which is definable only for a solid specimen with its own natural boundaries. (Under the conditions of defining and measuring the Young's modulus, the column of material is free to contract laterally when stretched, and to expand laterally when compressed, whereas with a gas we must provide a container with essentially rigid walls.) The appropriate modulus is that corresponding to changes of total *volume* of the specimen associated with a uniform stress in the form of a pressure change over its whole surface. This is the *bulk modulus*, K, referred to earlier; it is defined in general through the equation

$$
\begin{aligned}
K &= -\frac{dp}{dV/V} \\
&= -V\frac{dp}{dV}
\end{aligned}
\qquad (3\text{--}23)
$$

You will recall that Boyle's law describes the relation between pressure and volume for a gas at constant temperature. Thus Eq. (3–21) leads to a definition of the *isothermal* bulk modulus of a gas:

$$K_{\text{isothermal}} = p \qquad\qquad (3\text{–}24)$$

For a gas at atmospheric pressure this modulus is thus equal to about 10^5 N/m^2, i.e., five or six orders of magnitude less than for familiar solid materials (see Table 3–2).

An important question is whether the spring constant of an air column is indeed defined by the isothermal elasticity. In general this is not the case. When a gas is suddenly compressed it becomes warmer as a result of the work done on it; in other words, the particles composing it are moving faster, on the average. We have ignored this effect in using Boyle's law to calculate the change of pressure (and hence the restoring force) for a given change of length of the air column. Since, according to the kinetic theory of gases, the pressure is proportional to the mean-squared molecular speed, this heating results in a greater restoring force than we would otherwise have, and the elastic modulus of the gas column is larger than the value p predicted by Eq. (3–24). Experience bears out this conclusion. The pressure is changed by a factor greater than the inverse ratio of the volumes. Under completely adiabatic conditions (no flow of heat into or out of the gas) the pressure–volume relationship turns out to be the following[1]:

$$pV^\gamma = \text{const.} \quad \text{(adiabatic)} \qquad\qquad (3\text{–}25)$$

From this we have

$$\ln p + \gamma \ln V = \text{const.}$$

$$\frac{1}{p}\frac{dp}{dV} + \frac{\gamma}{V} = 0$$

$$K_{\text{adiabatic}} = -V\frac{dp}{dV} = \gamma p \qquad\qquad (3\text{–}26)$$

The value of the constant γ is close to 1.67 for monatomic gases, 1.40 for diatomic gases, and is less than 1.40 for all others (at normal room temperatures). This enhanced elasticity under adiabatic conditions then increases the frequency of any vibrations involving enclosed volumes of gas.

[1]The exact basis of Eq. (3–25) will be considered when we discuss the speed of sound in a gas.

"The spring of air"

So far we have treated springs as though they had no inertia, and acted purely as reservoirs of elastic potential energy. This, of course, is at best an approximation, and in some circumstances the inertia of the spring itself may play a dominant role. By way of approaching this question, let us consider the problem, beloved by textbook writers, of a body of mass m attached to a uniform spring of total mass M and spring constant k.[1] How does the period of oscillation differ from what it would be if the spring were massless? Even without doing any calculations we can predict that the period will be lengthened. But by how much?

A simple (and on the face of it reasonable) approach is to suppose that the various parts of the spring undergo displacements proportional to their distances from the fixed end, as indicated in Fig. 3–11 (and just as in a static extension, as shown in Fig. 3–4). We can then calculate the total kinetic energy of the spring at any instant when the extension of its far end has a displacement x.

Let the relaxed length of the spring be l, and let distance measured from the fixed end be s ($0 \le s \le l$). Consider an element of the spring lying between s and $s + ds$. Its mass is given by

$$dM = \frac{M}{l}\,ds$$

and its displacement is the fraction s/l of x. Thus the kinetic energy of this small element is given by

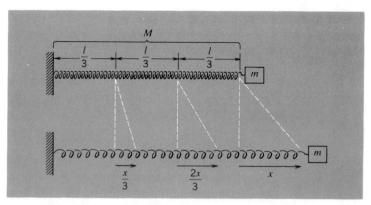

Fig. 3–11 Uniform extension of massive spring.

[1]As will appear later in this section, the problem has more than a merely pedantic interest if one considers it in the right context.

$$dK = \frac{1}{2}\left(\frac{M}{l}\,ds\right)\left(\frac{s}{l}\,\frac{dx}{dt}\right)^2$$

$$= \frac{M}{2l^3}\left(\frac{dx}{dt}\right)^2 s^2\,ds$$

At any given instant the total kinetic energy of the spring is obtained by integrating the above expression, treating dx/dt as a constant factor for this purpose. Hence we have

$$K_{\text{spring}} = \frac{M}{2l^3}\left(\frac{dx}{dt}\right)^2 \int_0^l s^2\,ds$$

$$= \tfrac{1}{6}M\left(\frac{dx}{dt}\right)^2$$

The energy-conservation statement for the whole system thus becomes

$$\tfrac{1}{2}m\left(\frac{dx}{dt}\right)^2 + \tfrac{1}{6}M\left(\frac{dx}{dt}\right)^2 + \tfrac{1}{2}kx^2 = E$$

giving

$$\omega^2 = \frac{k}{m + M/3}$$

It would be as if one took a massless spring and added $M/3$ to the mass attached to its end.

But is this true? Suppose, for example, that we took an extreme case in which we removed the attached mass m altogether, leaving ourselves with a system in which the spring itself was the repository of all the kinetic energy as well as all the elastic potential energy. Would the frequency of its free vibrations be given by $\omega = \sqrt{3k/M}$? The answer is no! The above calculation assumes the conditions of *static* extension of a uniform spring —an extension proportional to the distance from the fixed end. But this holds only if the stretching force is the same at all points along the spring. And if there is a distribution of mass along the spring, undergoing accelerations, this condition cannot possibly apply. There must be a variation of stretching force with distance along the spring. Our equation for ω is only an approximation; it is justified, however, if $M \ll m$, in which case the force along the spring *is* roughly constant (whereas, for $m = 0$, the restoring force must fall to zero at the free end, there being at this point an acceleration but no attached mass).

The above example, although imperfectly treated here, provides an important link between the simple mass–spring system and the free vibrations of an extended object. For, of course, a freely vibrating rod, or air column, is *precisely* like a massive spring with no mass attached at the end. It will be of central importance for us to analyze more exactly the behavior of such a system. We shall do this in Chapter 6. In the meantime, however, we can use the crude discussion above to suggest the *kind* of result that an exact treatment will give—that the frequency ν $(= \omega/2\pi)$ of a free oscillation of a uniform spring of mass M and spring constant k will be found to have the essential form

$$\nu = \text{const.} \sqrt{\frac{k}{M}} \tag{3-27}$$

(where the constant is a pure numerical factor), because this is the only combination of k and M that has the dimension of a frequency. We can even go a step further. Granted that Eq. (3-27) holds, we can substitute for k and M in terms of the linear dimensions, density, and elastic modulus of the material. Suppose, for example, that we have a solid rod, of length l, cross section A, density ρ, and Young's modulus Y. Then we have

$$M = Al$$
$$k = \frac{AY}{l} \quad \text{[Eq. (3-10)]}$$

and hence

$$\nu = \frac{\text{const.}}{l} \sqrt{\frac{Y}{\rho}} \tag{3-28}$$

We can expect such an equation to describe the longitudinal vibrations of a rod, although the numerical constant is as yet undetermined.

THE DECAY OF FREE VIBRATIONS

The free vibrations of any real physical system always die away with the passage of time. Every such system inevitably has dissipative features through which the mechanical energy of the vibration is depleted. Our very knowledge of the existence of a vibrating system is likely to imply a loss of energy on its part—as, for example, when we hear a tuning fork as the result of energy communicated by it to the air and then by the air to our ears. Thus it is never strictly correct to describe these free vibrations mathematically by a sinusoidal variation of constant amplitude. We shall now consider how the equation of free vibrations is modified by the introduction of dissipative forces.

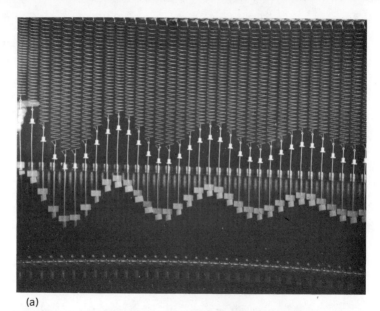

(a)

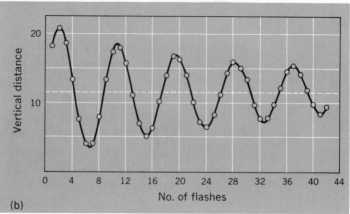

(b)

Fig. 3–12 (a) Multiple-flash photograph of free oscillations with damping. The camera was panned sideways to separate successive images. (Photo by Jon Rosenfeld, Education Research Center, M.I.T.). (b) Graph of damped oscillation, obtained by measuring a photograph obtained in this way.

We shall once again tie our discussion to the basic mass–spring system. Figure 3–12 shows an actual example of the decay of oscillations of such a system. To accentuate the damping, a vane attached to the moving mass was immersed in a cylinder of liquid; the multiple-flash photograph in Fig. 3–12(a) gives a clear picture of the course of the motion. Figure 3–12(b) is a graph based directly on measurements made on such a photograph.

The resistive force of a fluid to a moving object is some function of the velocity of the object; its magnitude is well described by the equation

$$R(v) = b_1 v + b_2 v^2$$

where v is the magnitude $|\mathbf{v}|$ of the velocity. This resistive force is exerted oppositely to the direction of $\mathbf{v}$ itself. Provided v is small compared to the ratio b_1/b_2, we can take the resistive force to be given by the linear term alone. In this case the statement of Newton's law for the moving mass can be written

$$m\frac{d^2x}{dt^2} = -kx - bv$$

i.e.,

$$m\frac{d^2x}{dt^2} + b\frac{dx}{dt} + kx = 0 \qquad (3\text{-}29)$$

or

$$\frac{d^2x}{dt^2} + \gamma\frac{dx}{dt} + \omega_0{}^2x = 0$$

where

$$\gamma = \frac{b}{m} \qquad \omega_0{}^2 = \frac{k}{m} \qquad (3\text{-}30)$$

It may be seen, then, that in this case the damping is characterized by the quantity γ, having the dimension of frequency, and the constant ω_0 would represent the angular frequency of the system if damping were absent.

Let us now seek a solution of Eq. (3-30). We shall do this by the complex exponential method, by assuming that x is the real part of a rotating vector z, where z satisfies an equation like Eq. (3-30), i.e.,

$$\frac{d^2z}{dt^2} + \gamma\frac{dz}{dt} + \omega_0{}^2z = 0 \qquad (3\text{-}30\text{a})$$

We shall assume a solution of the form

$$z = Ae^{j(pt+\alpha)} \qquad (3\text{-}31)$$

just like Eq. (3-8), and containing the requisite two constants, A and α, for the purpose of adjusting our solution to the initial values of displacement and velocity. Substituting in Eq. (3-30a) we find

$$(-p^2 + jp\gamma + \omega_0{}^2)Ae^{j(pt+\alpha)} = 0$$

If this is to be satisfied for all values of t, we must have

$$-p^2 + jp\gamma + \omega_0{}^2 = 0 \qquad (3\text{-}32)$$

This condition is one involving complex numbers; i.e., it really contains two conditions, applying to the real and imaginary components separately. It cannot be satisfied if the quantity p

is purely real, because the term $jp\gamma$ would then be a pure imaginary quantity with nothing to cancel it. We therefore put

$$p = n + js$$

where n and s are both real. Then

$$p^2 = n^2 + 2jns - s^2$$

Substituting these in Eq. (3–32) gives the following:

$$-n^2 - 2jns + s^2 + jn\gamma - s\gamma + \omega_0{}^2 = 0$$

We thus have two separate equations:

Real parts: $\quad -n^2 + s^2 - s\gamma + \omega_0{}^2 = 0$

Imaginary parts: $\quad -2ns + n\gamma = 0$

From the second of these we get

$$s = \frac{\gamma}{2}$$

Substituting $s = \gamma/2$ in the first equation then gives

$$n^2 = \omega_0{}^2 - \frac{\gamma^2}{4}$$

Now look back to Eq. (3–31). Writing p as a complex quantity $n + js$, we have

$$z = Ae^{j(nt+jst+\alpha)}$$
$$= Ae^{-st}e^{j(nt+\alpha)}$$

and hence

$$x = Ae^{-st}\cos(nt + \alpha)$$

Substituting the explicit values of n and s we thus find the following solution:

$$x = Ae^{-\gamma t/2}\cos(\omega t + \alpha) \qquad (3\text{–}33)$$

where

$$\omega^2 = \omega_0{}^2 - \frac{\gamma^2}{4} = \frac{k}{m} - \frac{b^2}{4m^2} \qquad (3\text{–}34)$$

Figure 3–13 shows a plot of Eq. (3–33) for the particular case $\alpha = 0$. The envelope of the damped oscillatory curve is also plotted in the figure.[1] The zeros of the curve are equally spaced with a separation of $\omega \, \Delta t = \pi$, and so are the successive maxima and minima, but the maxima and minima are only approximately

[1]The notation has been modified very slightly, writing A_0 instead of A to denote the amplitude of the motion at $t = 0$.

65　The decay of free vibrations

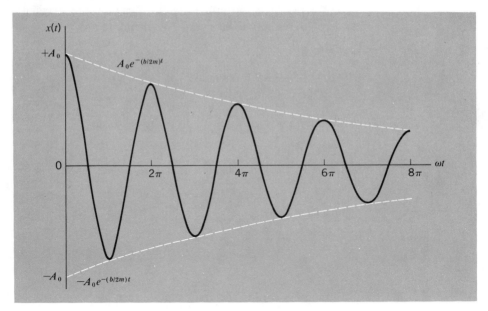

Fig. 3–13   Rapidly damped harmonic oscillations.

halfway between the zeros. Clearly ω may be identified as the natural angular frequency of the damped oscillator.

The curve in Fig. 3–13 is drawn for a case in which the decay of the vibrations is rapid. If, however, the damping is small, the motion approximates to SHM at constant amplitude over a number of cycles. Under these conditions, one can express the effect of the damping in terms of an exponential decay of the total mechanical energy, E. For, if $\gamma \ll \omega$, we can say that around time t the oscillations are well described over several cycles by SHM of constant amplitude A such that

$$A(t) = A_0 e^{-\gamma t/2} \tag{3–35}$$

Now the total mechanical energy of a simple harmonic oscillator is given by

$$E = \tfrac{1}{2}kA^2$$

Hence, using the above value of A, we have

$$E(t) = \tfrac{1}{2}kA_0{}^2 e^{-\gamma t}$$

i.e.,

$$E(t) = E_0 e^{-\gamma t} \tag{3–36}$$

This decay of the total energy is illustrated in Fig. 3–14.

66   The free vibrations of physical systems

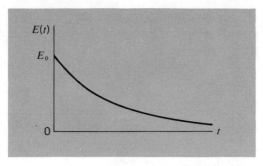

Fig. 3-14   Exponential decay of total energy during
the damping of harmonic oscillations.

You will recall that this whole analysis of the damping process
has been based on an assumption that the dissipation is due to a
resistive force proportional to the velocity. The situation would
be quite different (and far more difficult to handle) if some other
resistive law applied—e.g., $R(v) \sim v^2$. It is worth pointing out,
however, that the exponential decay of energy as described by
Eq. (3-35) may and does arise from many diverse kinds of dissipative processes. For example, in an oscillatory electrical circuit
the rate of energy dissipation in a resistor is proportional to the
square of the current, but so also is the total electric and magnetic
energy of the circuit. The situation is, in fact, closely analogous
to the mechanical oscillator with viscous damping.

In atomic and nuclear physics, also, there are many interactions that give rise to exponential decay of the energy of a
system and which lead to behavior of these systems analogous to
that of a simple mechanical oscillator with viscous damping.
Consequently, the analysis of such a mechanical oscillator provides one with some insight into all similar phenomena, although
it is a special case.

From the foregoing analysis, it is clear that the damped
oscillator is characterized by two parameters, ω_0 and γ ($= b/m$).
The constant ω_0 is the angular frequency of undamped oscillations
and γ is the reciprocal of the time required for the energy to
decrease to $1/e$ of its initial value. Thus ω_0 and γ are quantities
of the same dimensions. For convenience in applying our results
to diverse kinds of physical systems, we define a parameter called
the Q value (Q for *quality*) of the oscillatory system, given by
the ratio of these two quantities:

$$Q = \frac{\omega_0}{\gamma} \qquad (3\text{-}37)$$

Q is a pure number, large compared to unity for oscillating systems with small rates of dissipation of energy. In terms of the Q value, Eq. (3–34) becomes

$$\omega^2 = \omega_0{}^2 \left(1 - \frac{1}{4Q^2}\right)$$ (3–38)

If Q is large compared to unity, and this important case is the one with which we shall be mainly concerned, Eq. (3–38) gives $\omega \approx \omega_0$ and the motion of the oscillator [Eq. (3–33)] is given very nearly by

$$x = A_0 e^{-\omega_0 t/2Q} \cos(\omega_0 t + \alpha)$$ (3–39)

It may be noted that Q is closely related to the number of cycles of oscillation over which the amplitude of oscillation falls by a factor e. For according to Eq. (3–39) we have

$$A(t) = A_0 e^{-\omega_0 t/2Q}$$

Let us measure the time t in terms of the number of complete cycles of oscillation, n. Then, given the approximation that $\omega \approx \omega_0$, we can put $t \approx 2\pi n/\omega_0$. In terms of the number of cycles elapsed, therefore, we can put

$$A(n) \approx A_0 e^{-n\pi/Q}$$ (3–40)

so that the amplitude falls by a factor e in about Q/π cycles of free oscillation.

In terms of ω_0 and Q, we can rewrite Eq. (3–30) in the form

$$\frac{d^2 x}{dt^2} + \frac{\omega_0}{Q} \frac{dx}{dt} + \omega_0{}^2 x = 0$$ (3–41)

and this will in many cases be a highly convenient form of the basic differential equation for free oscillations, including damping, of a great variety of physical systems, both mechanical and nonmechanical.

THE EFFECTS OF VERY LARGE DAMPING[1]

You will have noted that the establishment of the equation for free damped oscillations [Eq. (3–33)] depends essentially upon our ability to introduce for these oscillations the angular frequency ω defined by the equation

$$\omega^2 = \omega_0{}^2 - \frac{\gamma^2}{4}$$

[1] Not strictly relevant to the oscillatory problem as such, but very closely connected and added for the sake of completeness.

But what if $\omega_0 \, [= (k/m)^{1/2}]$ is *less* than $\gamma/2 \, (= b/2m)$? In this case the motion is no longer oscillatory at all. We can get a strong hint as to the form of the solution to the problem by referring to the analysis preceding Eq. (3–33). We found that the differential equation of motion [Eq. (3–30)] is satisfied by a solution of the form

$$x = \text{Re}\,[Ae^{-\gamma t/2}e^{j(nt+\alpha)}]$$

where

$$n^2 = \omega_0{}^2 - \frac{\gamma^2}{4}$$

Suppose now that $\omega_0{}^2 < \gamma^2/4$. Then we can put

$$n^2 = -(\gamma^2/4 - \omega_0{}^2)$$

and if we proceed to solve for n we have

$$n = \pm j(\gamma^2/4 - \omega_0{}^2)^{1/2} = \pm j\beta, \qquad \text{say}$$

Thus we have $e^{jnt} = e^{\mp \beta t}$, which would define an exponential decay of x with t according to one or other of two possible exponents:

$$e^{-(\gamma/2+\beta)t} \qquad \text{or} \qquad e^{-(\gamma/2-\beta)t}$$

A rigorous analysis shows that both exponentials are in general necessary, and that the complete variation of x with t is given by the following equation:

$$x = A_1 e^{-(\gamma/2+\beta)t} + A_2 e^{-(\gamma/2-\beta)t} \tag{3–42}$$

where

$$\beta = \left(\frac{\gamma^2}{4} - \omega_0{}^2\right)^{1/2}$$

The two adjustable constants A_1 and A_2 (which may be of either sign) allow for the solution to be fitted to any given values of x and dx/dt at a given instant, e.g., $t = 0$.

One last question may be raised in connection with this heavily damped motion. What happens if ω_0 and $\gamma/2$ are *exactly equal* to one another? In this case the right side of Eq. (3–42) would reduce to two terms of exactly the same type, and only one adjustable constant would remain. This is not, however, an acceptable solution any longer; we still need two adjustable constants. It turns out that the appropriate form of solution for this case is

$$x = (A + Bt)e^{-\gamma t/2} \tag{3–43}$$

You can verify by substitution that this satisfies the basic equation of motion Eq. (3–30) if $\omega_0 = \gamma/2$ or $\gamma = 2\omega_0$ exactly. This very special condition corresponds to what is called *critical damping*. In real mechanical systems the value of the damping constant γ is often deliberately adjusted to meet this condition because, under conditions of critical damping, a constant force suddenly applied to the system (previously quiescent) will be followed by a smooth approach to a new, displaced position of equilibrium with no oscillation or overshoot. Such behavior is highly advantageous in the moving parts of electrical meters and the like, with which one may want to take a steady reading as soon as possible after the meter has been connected or a switch closed.

PROBLEMS

3–1   An object of mass 1 g is hung from a spring and set in oscillatory motion. At $t = 0$ the displacement is 43.785 cm and the acceleration is -1.7514 cm/sec^2. What is the spring constant?

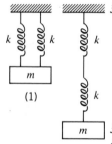

3–2   A mass m hangs from a uniform spring of spring constant k.
  (a) What is the period of oscillations in the system?
  (b) What would it be if the mass m were hung so that
     (1) It was attached to two identical springs hanging side by side?
     (2) It was attached to the lower of two identical springs connected end to end? (See Figure)

3–3   A platform is executing simple harmonic motion in a vertical direction with an amplitude of 5 cm and a frequency of $10/\pi$ vibrations per second. A block is placed on the platform at the lowest point of its path.
  (a) At what point will the block leave the platform?
  (b) How far will the block rise above the highest point reached by the platform?

3–4   A cylinder of diameter d floats with l of its length submerged. The total height is L. Assume no damping. At time $t = 0$ the cylinder is pushed down a distance B and released.
  (a) What is the frequency of oscillation?
  (b) Draw a graph of velocity versus time from $t = 0$ to $t =$ one period. The correct amplitude and phase should be included.

3–5   A uniform rod of length L is nailed to a post so that two thirds of its length is below the nail. What is the period of small oscillations of the rod?

3-6 A circular hoop of diameter d hangs on a nail. What is the period of its oscillations at small amplitude?

3-7 A wire of unstretched length l_0 is extended by a distance $10^{-3}l_0$ when a certain mass is hung from its bottom end. If this same wire is connected between two points, A and B, that are a distance l_0 apart on the same horizontal level, and the same mass is hung from the midpoint of the wire as shown, what is the depression y of the midpoint, and what is the tension in the wire?

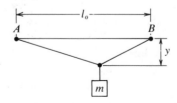

3-8 (a) An object of mass 0.5 kg is hung from the end of a steel wire of length 2 m and of diameter 0.5 mm. (Young's modulus = 2×10^{11} N/m^2). What is the extension of the wire?

(b) The object is lifted through a distance h (thus allowing the wire to become slack) and is then dropped so that the wire receives a sudden jerk. The ultimate strength of steel is 1.1×10^9 N/m^2. What is the largest possible value of h if the wire is not to break?

3-9 (a) A solid steel ball is to be hung at the bottom end of a steel wire of length 2 m and radius 1 mm. The ultimate strength of steel is 1.1×10^9 N/m^2. What are the radius and the mass of the biggest ball that the wire can bear?

(b) What is the period of torsional oscillation of this system? (Shear modulus of steel = 8×10^{10} N/m^2. Moment of inertia of sphere about axis through center = $2MR^2/5$.)

3-10 A metal rod, 0.5 m long, has a rectangular cross section of area 2 mm^2.

(a) With the rod vertical and a mass of 60 kg hung from the bottom, there is an extension of 0.25 mm. What is Young's modulus (N/m^2) for the material of the rod?

(b) The rod is firmly clamped at the bottom as shown in the sketch, and at the top a force F is applied in the y direction as shown (parallel to the edge of length b). The result is a static deflection, y, given by

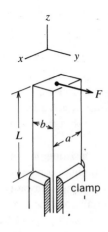

$$y = \frac{4L^3}{Yab^3} F$$

If the force F is removed and a mass m, which is much greater than the mass of the rod, is attached to the top end of the rod, what is the ratio of the frequencies of vibration in the y and x directions (i.e., parallel to edges of length b and a)?

y

x

(c) The mass is pulled aside in a certain transverse direction and released. It then traces a path like the one sketched. What is the ratio of a to b?

3-11 (a) Find the frequency of vibration under adiabatic conditions of a column of gas confined to a cylindrical tube, closed at one end, with a well-fitting but freely moving piston of mass m.

(b) A steel ball of diameter 2 cm oscillates vertically in a precision-bore glass tube mounted on a 12-liter flask containing air at atmospheric pressure. Verify that the period of oscillation should be about 1 sec. (Assume adiabatic pressure change with $\gamma = 1.4$. Density of steel $= 7600 \text{ kg/m}^3$.)

3-12 The motion of a linear oscillator may be represented by means of a graph in which x is shown as abscissa and dx/dt as ordinate. The history of the oscillator is then a curve.

(a) Show that for an undamped oscillator this curve is an ellipse.

(b) Show (at least qualitatively) that if a damping term is introduced one gets a curve spiraling into the origin.

3-13 Verify that $x = Ae^{-\alpha t} \cos \omega t$ is a possible solution of the equation

$$\frac{d^2x}{dt^2} + \gamma \frac{dx}{dt} + \omega_0^2 x = 0$$

and find α and ω in terms of γ and ω_0.

3-14 An object of mass 0.2 kg is hung from a spring whose spring constant is 80 N/m. The object is subject to a resistive force given by $-bv$, where v is its velocity in meters per second.

(a) Set up the differential equation of motion for free oscillations of the system.

(b) If the damped frequency is $\sqrt{3}/2$ of the undamped frequency, what is the value of the constant b?

(c) What is the Q of the system, and by what factor is the amplitude of the oscillation reduced after 10 complete cycles?

3-15 Many oscillatory systems, although the loss or dissipation mechanism is not analogous to viscous damping, show an exponential decrease in their stored *average* energy with time, $\bar{E} = \bar{E}_0 e^{-\gamma t}$. A Q for such oscillators may be defined using the definition $Q = \omega_0/\gamma$, where ω_0 is the natural angular frequency.

(a) When the note "middle C" on the piano is struck, its energy of oscillation decreases to one half its initial value in about 1 sec. The frequency of middle C is 256 Hz. What is the Q of the system?

(b) If the note an octave higher (512 Hz) takes about the same time for its energy to decay, what is its Q?

(c) A free, damped harmonic oscillator, consisting of a mass $m = 0.1$ kg moving in a viscous liquid of damping coefficient b ($F_{\text{viscous}} = -bv$), and attached to a spring of spring constant $k = 0.9$ N/m, is observed as it performs damped oscillatory motion.

Its average energy decays to $1/e$ of its initial value in 4 sec. What is the Q of the oscillator? What is the value of b?

3–16 According to classical electromagnetic theory an accelerated electron radiates energy at the rate Ke^2a^2/c^3, where $K = 6 \times 10^9$ N–m^2/C^2, e = electronic charge (C), a = instantaneous acceleration (m/sec^2), and c = speed of light (m/sec).

(a) If an electron were oscillating along a straight line with frequency ν (Hz) and amplitude A, how much energy would it radiate away during 1 cycle? (Assume that the motion is described adequately by $x = A \sin 2\pi\nu t$ during any one cycle.)

(b) What is the Q of this oscillator?

(c) How many periods of oscillation would elapse before the energy of the motion was down to half the initial value?

(d) Putting for ν a typical optical frequency (i.e., for visible light) estimate numerically the approximate Q and "half-life" of the radiating system.

3–17 A U-tube has vertical arms of radii r and $2r$, connected by a horizontal tube of length l whose radius increases linearly from r to $2r$. The U-tube contains liquid up to a height h in each arm. The liquid is set oscillating, and at a given instant the liquid in the narrower arm is a distance y above the equilibrium level.

(a) Show that the potential energy of the liquid is given by $U = \frac{5}{8}g\rho\pi r^2y^2$.

(b) Show that the kinetic energy of a small slice of liquid in the horizontal arm (see the diagram) is given by

$$dK = \tfrac{1}{2}\rho \, \frac{\pi r^2 \, dx}{(1 + x/l)^2}\left(\frac{dy}{dt}\right)^2$$

(Note that, if liquid is not to pile up anywhere, the product velocity $\times$ cross section must have the same value everywhere along the tube.)

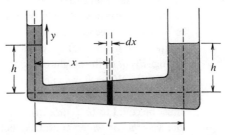

(c) Using the result of part (b), show that the total kinetic energy of all the moving liquid is given by

$$K = \tfrac{1}{4}\rho\pi r^2(l + \tfrac{5}{2}h)\left(\frac{dy}{dt}\right)^2$$

(Ignore any nastiness at the corners.)

(d) From (a) and (c), calculate period of oscillations if $l = 5h/2$.

3–18 This problem is much more ambitious than the usual problems, in the sense that it requires putting together a greater number of parts. But if you tackle the various parts as suggested, you should find that they are not, individually, especially difficult, and the problem as a whole exemplifies the power of the energy-conservation method for analyzing oscillation problems.

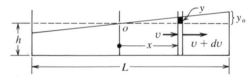

You are no doubt familiar with the phenomenon of water sloshing about in the bathtub. The simplest motion is, to some approximation, one in which the water surface just tilts as shown but seems to remain more or less flat. A similar phenomenon occurs in lakes and is called a seiche (pronounced: saysh). Imagine a lake of rectangular cross section, as shown, of length L and with water depth h ($\ll L$). The problem resembles that of the simple pendulum, in that the kinetic energy is almost entirely due to *horizontal* flow of the water, whereas the *potential* energy depends on the very small change of vertical level. Here is a program for calculating, *approximately*, the period of the oscillations:

(a) Imagine that at some instant the water level at the extreme ends is at $\pm y_0$ with respect to the normal level. Show that the increased gravitational potential energy of the whole mass of water is given by

$$U = \tfrac{1}{6}b\rho g L y_0{}^2$$

where b is the width of the lake. You get this result by finding the increased potential energy of a slice a distance x from the center and integrating.

(b) Assuming that the water flow is predominantly horizontal, its speed v must vary with x, being greatest at $x = 0$ and zero at $x = \pm L/2$. Because water is incompressible (more or less) we can relate the difference of flow velocities at x and $x + dx$ to the rate of change dy/dt of the height of the water surface at x. This is a *continuity* condition. Water flows in at x at the rate vhb and flows out at $x + dx$ at the rate $(v + dv)hb$. (We are assuming $y_0 \ll h$.) The difference must be equal to $(b\,dx)(dy/dt)$, which represents the rate of increase of the volume of water contained between x and $x + dx$. Using this condition, show that

$$v(x) = v(0) - \frac{1}{hL}x^2\frac{dy_0}{dt}$$

where

$$v(0) = \frac{L}{4h}\frac{dy_0}{dt}$$

(c) Hence show that at any given instant, the total kinetic energy associated with horizontal motion of the water is given by

$$K = \frac{1}{60} \frac{b\rho L^3}{h} \left(\frac{dy_0}{dt}\right)^2$$

To get this result, one must take the kinetic energy of the slice of water lying between x and $x + dx$ (with volume equal to $bh\,dx$), which moves with speed $v(x)$, and integrate between the limits $x = \pm L/2$.

(d) Now put

$$K + U = \text{const.}$$

This is an equation of the form

$$A\left(\frac{dy_0}{dt}\right)^2 + By_0^2 = \text{const.}$$

and defines SHM of a certain period. You will find that this period depends only on the length L, the depth h, and g. [*Note*: This theory is not really correct. The water surface is actually a piece of a sine wave, not a plane surface. But our formula is correct to better than 1%. (The true answer is $T = 2L/\sqrt{gh}$).]

(e) The Lake of Geneva can be approximated as a rectangular tank of water of length about 70 km and of mean depth about 150 m. The period of its seiche has been observed to be about 73 min. Compare this with your formula.

3–19 A mass m rests on a frictionless horizontal table and is connected to rigid supports via two identical springs each of relaxed length l_0 and spring constant k (see figure). *Each spring is stretched to a length l considerably greater than l_0.* Horizontal displacements of m from its equilibrium position are labeled x (along AB) and y (perpendicular to AB).

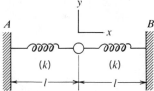

(a) Write down the differential equation of motion (i.e., Newton's law) governing small oscillations in the x direction.

(b) Write down the differential equation of motion governing small oscillations in the y direction (assume $y \ll l$).

(c) In terms of l and l_0, calculate the ratio of the *periods* of oscillation along x and y.

(d) If at $t = 0$ the mass m is released from the point $x = y = A_0$ with zero velocity, what are its x and y coordinates at any later time t?

(e) Draw a picture of the resulting path of m under the conditions of part (d) if $l = 9l_0/5$.

In the case of a cock putting its head into an empty utensil of glass where it crowed so that the utensil thereby broke, the whole cost shall be payable.

The Talmud (Baba Kamma, Chapter 2)

4

Forced vibrations and resonance

THE PRECEDING CHAPTER was concerned entirely with the free vibrations of various types of physical systems. We shall now turn to the remarkable phenomena, of profound importance throughout physics, that occur when such a system—a physical oscillator—is subjected to a periodic driving force by an external agency.

The key word is "resonance." Everybody has at least a qualitative familiarity with this phenomenon, and probably the most striking feature of a driven oscillator is the way in which a periodic force of a fixed size produces very different results depending on its frequency. In particular, if the driving frequency is made close to the natural frequency, then (as anyone who has pushed a swing knows) the amplitude of oscillation can be made very large by repeated applications of a quite small force. This is the phenomenon of resonance. A force of about the same size at frequencies well above or well below the resonant frequency is much less effective; the amplitude produced by it remains quite small. To judge by the quotation at the beginning of this chapter, the phenomenon has been recognized for a very long time.[1] It

[1]As Alexander Wood remarks in his book *Acoustics* (Blackie & Son, London, 1940): "It seems difficult to believe that legislation should be designed to cover a situation that had never arisen." The example does seem rather bizarre, however, and H. Bouasse, the French physicist who drew attention to this Talmudic pronouncement, reported that he had himself reared a large number of cocks, none of which developed a habit of putting their heads inside glass vases!

is typical of this type of motion that the driven system is compelled to accept whatever repetition frequency the driving force has; its tendency to vibrate at its own natural frequency may be in evidence at first, but ultimately gives way to the external influence.

To provide some initial feeling for the theoretical description of the resonance phenomenon, without getting too involved with analytical details, we shall begin by considering the simple though physically unreal case of an oscillator in which the damping effect is entirely negligible.

UNDAMPED OSCILLATOR WITH HARMONIC FORCING

We shall take our system to be the usual mass m on a spring of spring constant k. To this we shall imagine the application of a sinusoidal driving force $F = F_0 \cos \omega t$. The value of $\sqrt{k/m}$, representing the natural angular frequency of the system, will be denoted by ω_0. Then the statement of the equation of motion, in the form $ma = $ net force, is

$$m \frac{d^2x}{dt^2} = -kx + F_0 \cos \omega t$$

or

$$m \frac{d^2x}{dt^2} + kx = F_0 \cos \omega t \qquad (4\text{-}1)$$

Before we discuss this differential equation of motion in detail, let us consider the situation qualitatively. If the oscillator is driven from its equilibrium position and then left to itself, it will oscillate with its natural frequency ω_0. A periodic driving force will, however, try to impose its own frequency[1] ω on the oscillator. We must expect, therefore, that the actual motion in this case is some kind of a superposition of oscillations at the two frequencies ω and ω_0. The mathematically complete solution of Eq. (4-1) is indeed a simple sum of these two motions. But because of the inevitable presence of dissipative forces in any real system, the free oscillations will eventually die out. The initial stage, in which the two types of motion are both prominent, is called the *transient*. After a sufficiently long time, however, the only motion in effect present is the forced oscillation, which will continue undiminished at the frequency ω. When this condition has been achieved, we

[1] To avoid tiresome repetitions, we shall often refer to ω simply as "frequency" rather than "angular frequency" in contexts where no ambiguity is entailed.

have what is called a steady-state motion of the driven oscillator.

Later we shall analyze the transient effects, but for the present we shall focus our attention exclusively on the steady state of the forced oscillation. In an ideal undamped oscillator, the effect of the natural vibrations would never disappear, but we shall temporarily ignore this embarrassing fact for the sake of the simplicity that absence of damping brings to the forced-motion problem.

The most striking feature of the motion will be the large response near $\omega = \omega_0$, but before embarking on the solution of Eq. (4–1) in its entirety, let us point to some features of the motion in the extremes of very low or very high values of the driving frequency ω. If the driving force is of very low frequency relative to the natural frequency of free oscillations, we would expect the particle to move essentially in step with the driving force with an amplitude not very different from F_0/k $(= F_0/m\omega_0{}^2)$, the displacement which a constant force F_0 would produce. This is equivalent to stating that the term $m(d^2x/dt^2)$ in Eq. (4–1) plays a relatively small role compared to the term kx at very low frequencies, or in other words that the response is controlled by the stiffness of the spring. On the other hand, at frequencies of the driving force very large compared to the natural frequency of free oscillation, the opposite situation holds. The term kx becomes small compared to $m(d^2x/dt^2)$ because of the large acceleration associated with high frequencies, so that the response is controlled by the inertia. In this case we expect a relatively small amplitude of oscillation and this oscillation should be opposite in phase to the driving force, because the acceleration of a particle in harmonic motion is 180° out of phase with its displacement. It is still not apparent from these remarks that the resonant amplitude should greatly exceed that at low or high frequencies, but this we shall now show.

To obtain the steady-state solution of Eq. (4–1) we set

$$x = C \cos \omega t \qquad\qquad (4\text{–}2)$$

We are assuming, in other words, that the motion is harmonic, of the same frequency and phase as the driving force, and that the natural oscillations of the system are not present. It must be kept in mind that the assumption of Eq. (4–2) is tentative and we must be prepared to reject it if we fail to find a value of the as-yet-undetermined constant C such that Eq. (4–1) is satisfied for arbitrary values of ω and t. Differentiating Eq. (4–2) twice

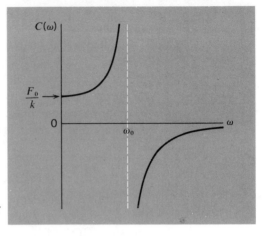

Fig. 4-1 Amplitude of forced oscillations as a function of the driving frequency (assuming zero damping.) The negative sign of the amplitude for $\omega > \omega_0$ corresponds to a phase lag π of displacement with respect to driving force.

with respect to t, we get

$$\frac{d^2x}{dt^2} = -\omega^2 C \cos \omega t$$

Substituting in Eq. (4-1) we thus have

$$-m\omega^2 C \cos \omega t + kC \cos \omega t = F_0 \cos \omega t$$

and hence

$$C = \frac{F_0}{k - m\omega^2} = \frac{F_0/m}{\omega_0^2 - \omega^2} \tag{4-3}$$

Equation (4-3) satisfactorily defines C in such a way that Eq. (4-1) is always satisfied. Thus we can take it that the forced motion is indeed described by Eq. (4-2), with C depending on ω according to Eq. (4-3). This dependence is shown graphically in Fig. 4-1. Notice how C switches abruptly from large positive to large negative values as ω passes through ω_0. The resonance phenomenon itself is represented by the result that the magnitude of C, without regard to sign, becomes infinitely large at $\omega = \omega_0$ exactly.

Although Eqs. (4-2) and (4-3) between them describe in a perfectly adequate way the solution of this dynamical problem, there is a better way of stating the result, more in accord with our general description of harmonic motions. This is to express x in terms of a sinusoidal vibration having an amplitude A, by definition a *positive* quantity, and a phase α at $t = 0$.

$$x = A \cos(\omega t + \alpha) \tag{4-4}$$

It is not difficult to see that this implies putting $A = |C|$ and giving α one or other of two values, according to whether the driving frequency ω is less or greater than ω_0:

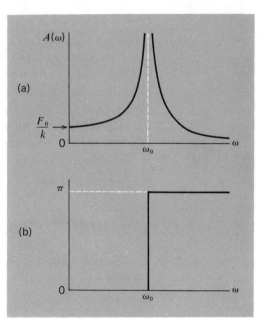

Fig 4–2 (a) Absolute amplitude of forced oscillations as a function of the driving frequency, for zero damping.
(b) Phase lag of the displacement with respect to the driving force as a function of frequency.

$$\omega < \omega_0: \alpha = 0$$
$$\omega > \omega_0: \alpha = \pi$$

The response of the system over the whole range of ω is then represented by separate curves for the amplitude A and the phase α, as shown in Fig. 4–2. The infinite value of A at $\omega = \omega_0$, and the discontinuous jump from zero to π in the value of α as one passes through ω_0, must be unphysical, but, as we shall see, they represent a mathematically limiting case of what actually occurs in systems with nonzero damping.

The actual reversal of phase of the displacement with respect to the driving force (i.e., from being in phase to being 180° out

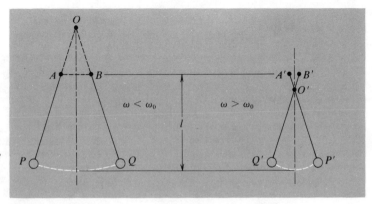

Fig. 4–3 Motion of simple pendulums resulting from forced harmonic oscillation of the point of suspension along the line AB. (a) $\omega < \omega_0$. (b) $\omega > \omega_0$.

of phase) is shown in a very direct way by the behavior of a simple pendulum that is driven by moving its point of suspension back and forth horizontally in SHM. The situations for frequencies well below and well above resonance are illustrated in Fig. 4–3. Once the steady state has been established, the pendulum behaves as though it were suspended from a fixed point corresponding to a length greater than its true length l for $\omega < \omega_0$, and less than l for $\omega > \omega_0$. In the former case the motion of the bob is always in the same direction as the motion of the suspension, whereas in the latter case it is always opposite.

THE COMPLEX EXPONENTIAL METHOD FOR FORCED OSCILLATIONS

Having dealt with this simplest of forced vibration problems in terms of sinusoidal functions, let us do it again using the complex exponential. This has no special merit as far as the present problem is concerned, but the technique, illustrated here in elementary terms, will show to great advantage when we come to deal with the damped oscillator. Our program is as follows:

1. We start with the physical equation of motion as given by Eq. (4–1):

$$m\frac{d^2x}{dt^2} + kx = F_0 \cos \omega t$$

2. We imagine the driving force $F_0 \cos \omega t$ as being the projection on the x axis of a rotating vector $F_0 \exp(j\omega t)$, as shown in Fig. 4–4(a), and we imagine x as being the projection of a vector z that rotates at the same frequency ω [Fig. 4–4(b)].

3. We then write the differential equation that governs z:

Fig. 4–4 (a) Complex representation of sinusoidal driving force. (b) Complex representation of displacement vector in the forced oscillation.

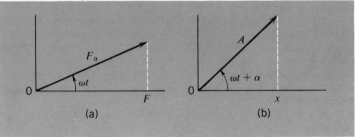

(a)   (b)

$$m\frac{d^2z}{dt^2} + kz = F_0 e^{j\omega t} \tag{4-5}$$

4. We try the solution

$$z = Ae^{j(\omega t + \alpha)}$$

Substituting in Eq. (4-5) this gives us

$$(-m\omega^2 A + kA)e^{j(\omega t + \alpha)} = F_0 e^{j\omega t}$$

which can be rewritten as follows:

$$(\omega_0^2 - \omega^2)A = \frac{F_0}{m}e^{-j\alpha}$$

$$= \frac{F_0}{m}\cos\alpha - j\frac{F_0}{m}\sin\alpha \tag{4-6}$$

This contains two conditions, corresponding to the real and imaginary parts on the two sides of the equation:

$$(\omega_0^2 - \omega^2)A = \frac{F_0}{m}\cos\alpha$$

$$0 = -\frac{F_0}{m}\sin\alpha$$

These clearly lead at once to the solutions represented by the two graphs in Fig. 4-2.

FORCED OSCILLATIONS WITH DAMPING

At the end of Chapter 3 we analyzed the free vibrations of a mass–spring system subject to a resistive force proportional to velocity. We shall now consider the result of acting on such a system with a force just like that considered in the previous section. The statement of Newton's law then becomes

$$m\frac{d^2x}{dt^2} = -kx - b\frac{dx}{dt} + F_0 \cos\omega t$$

or

$$\frac{d^2x}{dt^2} + \frac{b}{m}\frac{dx}{dt} + \frac{k}{m}x = \frac{F_0}{m}\cos\omega t$$

Putting $k/m = \omega_0^2$, $b/m = \gamma$, this can be written

$$\frac{d^2x}{dt^2} + \gamma\frac{dx}{dt} + \omega_0^2 x = \frac{F_0}{m}\cos\omega t \tag{4-7}$$

Let us now look for a steady-state solution to this equation.

We shall go at once to the complex-exponential method; our basic equation then becomes the following:

$$\frac{d^2z}{dt^2} + \gamma\frac{dz}{dt} + \omega_0{}^2 z = \frac{F_0}{m}e^{j\omega t} \tag{4-8}$$

We shall now assume the following solution:

$$z = Ae^{j(\omega t - \delta)} \tag{4-9}$$

with

$$x = \text{Re}\,(z)$$

Notice that we have assumed a slightly different equation for z than we did in the previous section; we have written the initial phase of z as $-\delta$ instead of $+\alpha$. Why did we do this? The clue is to be found in Eq. (4-6). The right-hand side of the equation can be read, in geometrical terms, as an instruction to take a vector of length F_0/m and rotate it through the angle $-\alpha$ with respect to the real axis. We are going to get a very similar equation now, and it will simplify things if we define our angle, formally at least, as representing a positive (counterclockwise) rotation. That is, δ is formally a positive phase angle by which the driving force leads the displacement.

Substituting from Eq. (4-9) into Eq. (4-8) we thus get

$$(-\omega^2 A + j\gamma\omega A + \omega_0{}^2 A)e^{j(\omega t - \delta)} = \frac{F_0}{m}e^{j\omega t}$$

Therefore,

$$(\omega_0{}^2 - \omega^2)A + j\gamma\omega A = \frac{F_0}{m}e^{j\delta} \tag{4-10}$$

Now the elegance and perspicuity of the complex exponential method are really displayed. We can read Eq. (4-10) as a geometrical statement. The left-hand side tells us to draw a vector of length $(\omega_0{}^2 - \omega^2)A$, and then at right angles to it a vector of

Fig. 4-5 *Geometrical representation of Eq. (4-10).*

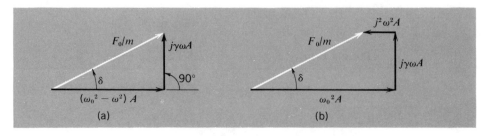

(a)

(b)

Forced vibrations and resonance

length $\gamma \omega A$. The right-hand side tells us to draw a vector of length F_0/m at an angle δ to the real axis. The equation requires that these two operations bring us to the same point, so that the vectors form a closed triangle, as shown in Fig. 4–5(a).[1] Clearly, we have

$$(\omega_0{}^2 - \omega^2)A = \frac{F_0}{m}\cos\delta$$

$$\gamma\omega A = \frac{F_0}{m}\sin\delta$$

Therefore,

$$A(\omega) = \frac{F_0/m}{[(\omega_0{}^2 - \omega^2)^2 + (\gamma\omega)^2]^{1/2}}$$

$$\tan\delta(\omega) = \frac{\gamma\omega}{\omega_0{}^2 - \omega^2}$$

(4–11)

These same results can of course be obtained without introducing complex exponentials. One simply assumes a solution of the form

$$x = A\cos(\omega t - \delta) \tag{4–12}$$

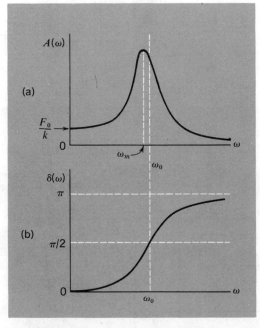

Fig. 4–6 (a) Dependence of amplitude upon driving frequency for forced oscillations with damping. (b) Phase of displacement with respect to driving force as a function of the driving frequency.

[1]You may actually prefer to read the left-hand side of Eq. (4–10) even more literally (in terms of its origins) as a sum of *three* vectors,

$$\omega_0{}^2 A + j\gamma\omega A + (j)^2\omega^2 A$$

as shown in Fig. 4–5(b).

and substitutes this in Eq. (4–7), which leads to the equation

$$(\omega_0{}^2 - \omega^2)A\cos(\omega t - \delta) - \gamma\omega A\sin(\omega t - \delta) = \frac{F_0}{m}\cos\omega t$$

This must then be solved as a trigonometric identity true for all t. The analysis is certainly not difficult, but it is less transparent

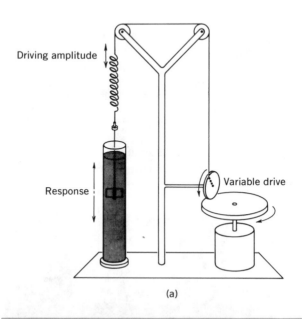

(a)

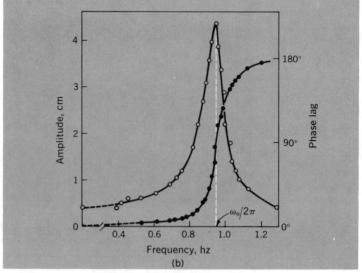

Fig. 4–7 (a) Di-
agrammatic sketch of
the "Texas Tower," a
mechanical resonance
apparatus developed
by J. G. King at the
Education Research
Center, M.I.T. (b)
Experimental
resonance curves for
amplitude and phase
lag obtained with this
apparatus. (Measure-
ments by G. J.
Churinoff, M.I.T.
class of 1967.)

and instructive than the other.

The type of dependence of amplitude A and phase angle δ upon frequency ω, for an assumed constant magnitude of F_0, is shown in Fig. 4–6. (Remember that δ is the angle by which the driving force leads the displacement, or by which the displacement lags behind the driving force.) These curves have a clear general resemblance to those in Fig. 4–2 for the undamped oscillator. As can be seen from the expression for $\tan \delta$ in equations (4–11), the phase lag increases continuously from zero (at $\omega = 0$) to 180° (in the limit $\omega \to \infty$); it passes through 90° at *precisely* the frequency ω_0. Less obvious is the fact that the maximum amplitude is attained at a frequency ω_m somewhat less than ω_0; in most cases of any practical interest, however, the difference between ω_m and ω_0 is negligibly small.

These are some of the calculated features of a forced, damped oscillator. How nearly are they exhibited by actual physical systems? Figure 4–7 provides an answer in the form of experimental results obtained with the type of physical system we have been discussing. It is, to be sure, not a natural system but an artificial one, devised specifically to display these features. Nevertheless, there is satisfaction in seeing that the pattern of behavior described by our mathematical analysis (which might, after all, bear no relation to reality) does, in fact, correspond quite well to the behavior of a system containing a real spring and a real viscous damping agency. This is the same system for which we showed the decay of free oscillations in Fig. 3–12.

The features of Fig. 4–6 can also be nicely demonstrated in a simple but, as it were, backhanded way, by applying a driving force of some *fixed* frequency to a whole collection of oscillators of different natural frequencies. This is readily done by a modification of an arrangement due to E. H. Barton (1918) in which a number of light pendulums of different lengths are hung from a horizontal bar that is rocked at the resonance frequency of one pendulum in the middle of the range, as shown in Fig. 4–8(a). When photographed edgewise the motions of the light pendulum bobs, all driven at the same frequency, display, qualitatively at least, the expected phase relationships. This is indicated in Fig. 4–8(b), which shows the displacements of the small pendulums at the instant when the driving bar is passing from left to right through its equilibrium position, and then at a slightly later instant. The short pendulums (for which $\omega_0 > \omega$) have

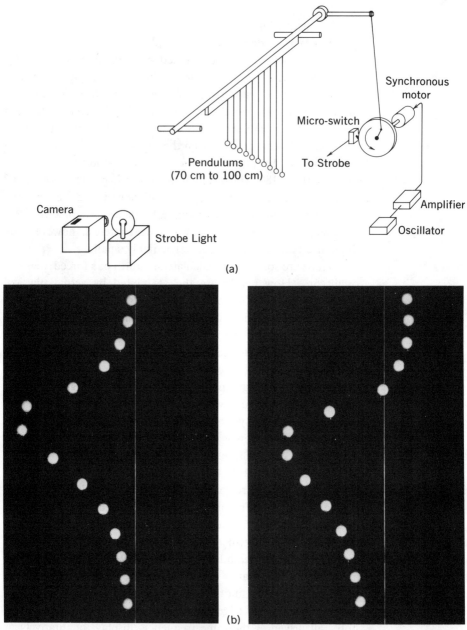

Fig. 4–8   *A modern version of Barton's pendulums experiment.   (a) A general sketch of the arrangement.   The strobe light flashes once per oscillation at a controllable point in the cycle.   (b) Displacements of the pendulums when the driving force is passing through zero (left) and at a somewhat later instant (right).   In the latter photograph, note that the shorter pendulums have moved in the same direction as the driver and the longer pendulums have moved in the opposite direction, corresponding to δ < 90° and δ > 90° respectively.   (Photos by Jon Rosenfeld, Education Research Center, M.I.T.).*

$\delta < 90°$, the long ones (for which $\omega_0 < \omega$) have $\delta > 90°$, and so move contrary to the driver, and the pendulum in exact resonance lags by $90°$, being at maximum negative displacement as the driver passes through zero.

EFFECT OF VARYING THE RESISTIVE TERM

In discussing the decay of free vibrations at the end of Chapter 3, we introduced the "quality factor" Q, the pure number equal to the ratio ω_0/γ. The larger the value of Q, the less the dissipative effect and the greater the number of cycles of free oscillation for a given decrease of amplitude. We shall now indicate how the behavior of the resonant system changes as the Q of the system is changed, other things being equal.

We shall put Eq. (4–11) (for A and $\tan \delta$) into more convenient form for this purpose. First, substituting $\gamma = \omega_0/Q$ gives us

$$A(\omega) = \frac{F_0/m}{[(\omega_0{}^2 - \omega^2)^2 + (\omega\omega_0/Q)^2]^{1/2}}$$

$$\tan \delta(\omega) = \frac{\omega\omega_0/Q}{\omega_0{}^2 - \omega^2}$$

(4–13)

Furthermore, it will prove convenient for many purposes to use the ratio ω/ω_0, rather than ω itself, as a variable. With this in mind we shall rewrite equations (4–13) in the following form:

$$A = \frac{F_0}{m\omega_0{}^2} \frac{\omega_0/\omega}{\left[\left(\dfrac{\omega_0}{\omega} - \dfrac{\omega}{\omega_0}\right)^2 + \dfrac{1}{Q^2}\right]^{1/2}}$$

or

$$A = \frac{F_0}{k} \frac{\omega_0/\omega}{\left[\left(\dfrac{\omega_0}{\omega} - \dfrac{\omega}{\omega_0}\right)^2 + \dfrac{1}{Q^2}\right]^{1/2}}$$

(4–14)

and

$$\tan \delta = \frac{1/Q}{\dfrac{\omega_0}{\omega} - \dfrac{\omega}{\omega_0}}$$

In Fig. 4–9 we show curves calculated from equations (4–14) to show the variations with frequency of amplitude A and phase lag δ for different values of Q. Most of the change of δ takes place over a range of frequencies roughly from $\omega_0(1 - 1/Q)$ to $\omega_0(1 + 1/Q)$, i.e., a band of width $2\omega_0/Q$ centered on ω_0. In the limit $Q \to \infty$ the phase lag jumps abruptly from zero to π as

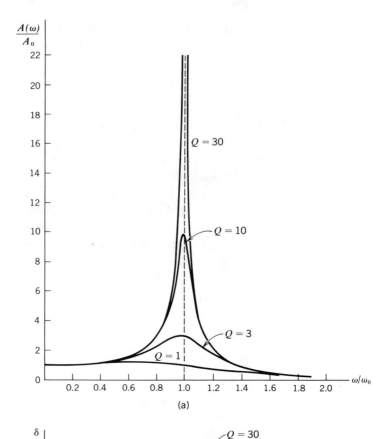

(a)

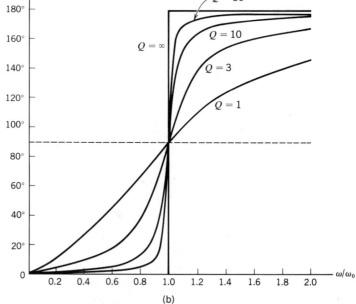

Fig. 4–9 (a) Amplitude as function of driving frequency for different values of Q, assuming driving force of constant magnitude but variable frequency. (b) Phase difference δ as function of driving frequency for different values of Q.

(b)

one passes through ω_0. Clearly the frequency ω_0 is an important property of the resonant system, even though it is not (except for zero damping) the frequency with which the system would oscillate when left to itself.

The amplitude A passes through a maximum for any value of Q greater than $1/\sqrt{2}$—i.e., for all except the most heavily damped systems. This maximum amplitude A_m occurs, as we noted earlier, at a frequency ω_m that is less than ω_0. If we denote by A_0 the amplitude F_0/k obtained for $\omega \to 0$, then one can readily show that the following results hold:

$$\omega_m = \omega_0 \left(1 - \frac{1}{2Q^2} \right)^{1/2}$$

$$A_m = A_0 \frac{Q}{\left(1 - \frac{1}{4Q^2} \right)^{1/2}}$$

(4-15)

In Table 4–1 we list some values of ω_m/ω_0 and A_m/A_0 for particular Q values. Notice that in most cases ($Q \geq 5$) the peak

TABLE 4–1:   RESONANCE PARAMETERS OF DAMPED SYSTEMS

Q	ω_m/ω_0	A_m/A_0
$1/\sqrt{2}$	0	1
1	$1/\sqrt{2} = 0.707$	$2/\sqrt{3} = 1.15$
2	$\sqrt{\tfrac{7}{8}} = 0.935$	$8/\sqrt{14} = 2.06$
3	$\sqrt{\tfrac{17}{18}} = 0.973$	$18/\sqrt{35} = 3.04$
5	$\sqrt{\tfrac{49}{50}} = 0.990$	$50/\sqrt{99} = 5.03$
$\gg 1$	$1 - 1/4Q^2$	$Q[1 + 1/(8Q^2)]$

amplitude is close to being Q times the static displacement for the same F_0, and it occurs at a frequency quite close to ω_0. At the frequency ω_0 itself the amplitude is precisely QA_0.

Figure 4–9 demonstrates how the sharpness of tuning of a resonant system varies with Q. The arrangement of an array of pendulums, as in Fig. 4–8(a), can be used to display the phenomenon. The Q can be increased, without changing ω_0, by making the bobs of the driven pendulums more massive. Figure 4–10 shows time-exposure photographs of the pendulums, first unloaded and then with two different degrees of loading. This clearly reveals the improvement in sharpness of tuning, even though the absolute amplitudes of oscillation in the three pictures are not strictly comparable. An instantaneous flash photograph is superimposed on each time-exposure photograph, displaying

91   Effect of varying the resistive term

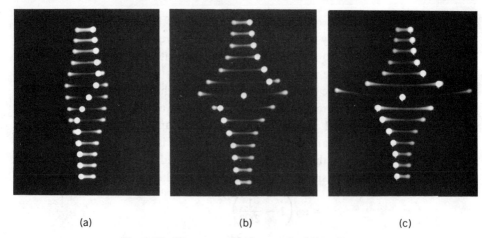

Fig. 4–10   Time exposure photograph of Barton's pendulums (cf. Fig. 4–8) showing resonance properties. The pendulum bobs were light styrofoam spheres (from PSSC Electrostatics Kit). (a) Pendulum bobs unloaded and therefore heavily damped, showing little selective resonance. (b) Each pendulum bob lightly loaded (with one thumbtack) giving moderate damping and more selective resonance. (c) Each pendulum bob heavily loaded (one thumbtack + one small washer) giving small damping and fairly high Q. (Photos by Jon Rosenfeld, Education Research Center, M.I.T.)   In each case an instantaneous flash photograph is superimposed in order to display the phase relationships among the driven pendulums.

the phase relationships among the driven pendulums for different Q, corresponding to Fig. 4–9(b).

TRANSIENT PHENOMENA

Our discussion so far has taken the steady state as being completely established, as if the driving force $F_0 \cos \omega t$ had been acting since far back in the past and all trace of any natural vibrations of the driven system had vanished. But of course in any real situation the driving force is first brought into action at some instant—which failing any reason to the contrary we might as well call $t = 0$—and it is only some time later that our steady-state conditions supervene. This transient stage may occupy a very long time indeed if the damping of the free vibrations is extremely small, and we shall even begin (again because of its mathematical simplicity) with the case in which the damping is effectively zero.

To make the problem quite explicit, let us suppose that we have a mass–spring system which, up to $t = 0$, is at rest. At $t = 0$ the driving force is turned on, and thereafter the motion is governed by Eq. (4–1), which we introduced at the beginning of this chapter:

$$m\frac{d^2x}{dt^2} + kx = F_0 \cos \omega t$$

or

$$\frac{d^2x}{dt^2} + \omega_0^2 x = \frac{F_0}{m} \cos \omega t \tag{4–16}$$

Now we have already seen how this differential equation of the forced motion leads to the following equation for x:

$$x = \frac{F_0/m}{\omega_0^2 - \omega^2} \cos \omega t \tag{4–17}$$

This equation, however, contains no adjustable constants of integration; the solution is completely specified by the values of m, ω_0, F_0, and ω. After our remarks in Chapter 3 about the need to introduce two constants of integration in solving a second-order differential equation, you may have wondered what became of them in this case. More specifically and, as it were, empirically, we can look at what Eq. (4–17) would give us for $t = 0$, the instant at which, according to our present assumptions, the driving force is first switched on. The result is impossible! If, for example, we suppose $\omega < \omega_0$, the displacement at $t = 0$ immediately assumes a positive value. But no system with nonzero inertia, acted on by a finite force, can be displaced through a nonzero distance in zero time. And if we suppose $\omega > \omega_0$, the result is a still greater absurdity—the mass would suddenly move to a *negative* displacement under the action of a *positive* force. Quite clearly Eq. (4–17) does not tell the whole story, and it is the transient that comes to the rescue.

Mathematically, the situation is this. Suppose that we have found a solution—call it x_1—to Eq. (4–16) so that

$$\frac{d^2x_1}{dt^2} + \omega_0^2 x_1 = \frac{F_0}{m} \cos \omega t$$

And now suppose that we have also found a solution—call it x_2— to the equation of free vibration, so that

$$\frac{d^2x_2}{dt^2} + \omega_0^2 x_2 = 0$$

Then by simple addition of these two equations we have

$$\frac{d^2(x_1 + x_2)}{dt^2} + \omega_0{}^2(x_1 + x_2) = \frac{F_0}{m}\cos \omega t$$

Thus the combination $x_1 + x_2$ is just as much a solution of the equation of forced motion as is x_1 alone. We have no mathematical reason to exclude the contribution from x_2; on the contrary, we are absolutely obliged to include it if we are to take care of the conditions existing at $t = 0$. We can say much the same thing, although less precisely, from a purely physical standpoint. The oscillations resulting from a brief impulse given to the system at $t = 0$ would certainly possess the natural frequency ω_0. It is only if a periodic force is applied over many cycles that the system learns, as it were, that it should oscillate with some different frequency ω. Thus one should expect that the motion, at least in its initial stages, contains contributions from both frequencies.

Turning now to the precise equations, the equation of the free vibration of frequency ω_0 *does* contain two adjustable constants—an amplitude and an initial phase. Let us call them B and β because we are using them to fit conditions at the beginning of the forced motion. Then, according to the ideas outlined above, we propose that the complete solution of the forced-motion equation is as follows:

$$x = B\cos(\omega_0 t + \beta) + C\cos \omega t \tag{4–18}$$

where

$$C = \frac{F_0/m}{\omega_0{}^2 - \omega^2}$$

We can now tailor Eq. (4–18) to fit the initial conditions (in this case) that $x = 0$ and $dx/dt = 0$ at $t = 0$. For the condition on x itself we have

$$0 = B\cos \beta + C$$

Also, differentiating Eq. (4–18), we have

$$\frac{dx}{dt} = -\omega_0 B \sin(\omega_0 t + \beta) - \omega C \sin \omega t$$

Hence, at $t = 0$, we have

$$0 = -\omega_0 B \sin \beta$$

The second condition requires that $\beta = 0$ or π. Taking the former (the final result is the same in either case) we get $B = -C$, so that Eq. (4–18) becomes

$$x = C(\cos \omega t - \cos \omega_0 t) \qquad (4\text{-}19)$$

which is a typical example of beats, as shown in Fig. 4-11(a). In the complete absence of damping these beats would continue indefinitely; no steady state corresponding to Eq. (4-17) alone would ever be reached. It is perhaps worth noting that the conditions just after $t = 0$ now make excellent sense. If ωt, $\omega_0 t \ll 1$, we can put

$$\cos \omega t \approx 1 - \frac{\omega^2 t^2}{2}$$

$$\cos \omega_0 t \approx 1 - \frac{\omega_0^2 t^2}{2}$$

Therefore,

$$x \approx \frac{F_0/m}{\omega_0^2 - \omega^2} \frac{(\omega_0^2 - \omega^2)t^2}{2} = \frac{1}{2}\frac{F_0}{m}t^2$$

Thus, precisely as we should expect, before the restoring forces have been called into play the mass starts out in the direction of the applied force with acceleration F_0/m.

You may wonder whether, granted that Eq. (4-18) can be justified as *a* solution of the forced-motion equation, it is therefore *the* solution. Here we shall merely assert that there is a uniqueness theorem for such differential equations, and if we have found any solution with the requisite number of adjustable constants, it is indeed the only solution of the problem.[1]

Turning now to the more realistic case in which damping is assumed to be present, we can without more ado postulate the following combination of free and steady-state motions:

$$x = Be^{-\gamma t/2}\cos(\omega_1 t + \beta) + A\cos(\omega t - \delta) \qquad (4\text{-}20)$$

where

$$\omega_1 = \left(\omega_0^2 - \frac{\gamma^2}{4}\right)^{1/2}$$

and A, δ are given by Eq. (4-11).

We shall not attempt here to delve into the purely mathematical details of fitting the values of B and β to the values of x and dx/dt at $t = 0$. It is just a more complicated version of what we did above for the undamped oscillator. In Fig. 4-11(b), however, we show the kind of motion that occurs—in general

[1]For a fuller discussion see, for example, W. T. Martin and E. Reissner, *Elementary Differential Equations*, Addison-Wesley, Reading, Mass., 2nd ed., 1961.

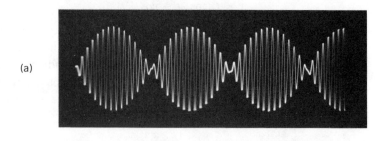

(a)

(b)

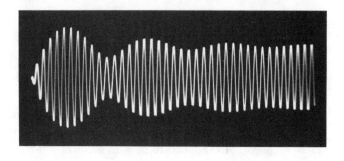

Fig. 4–11 (a) Response of an undamped harmonic oscillator to a periodic driving force, as described by Eq. (4–19). This beat pattern would continue indefinitely. (b) Transient behavior of a damped oscillator with a periodic driving force off resonance. (c) Transient behavior at exact resonance, showing smooth growth toward steady amplitude. (Photos by Jon Rosenfeld, Education Research Center, M.I.T.)

(c)

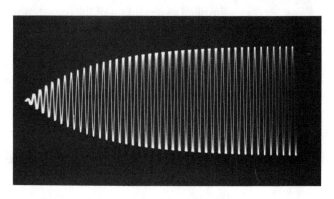

what looks like an attempt at beats, settling down to a motion of constant amplitude at the driving frequency ω. Figure 4–11(c) shows the much simpler transient effect that occurs when the damped oscillator is driven at its own natural frequency.

THE POWER ABSORBED BY A DRIVEN OSCILLATOR

It will often be a matter of importance and interest to know at what rate energy must be fed into a driven oscillator to maintain its oscillations at a fixed amplitude. As in any other dynamical situation, we can calculate the instantaneous power input, P, as the driving force times the velocity:

$$P = \frac{dW}{dt} = F\frac{dx}{dt} = Fv$$

velocity takes up both direction ↓↑

Once again, let us consider first the undamped oscillator, for which (because there are no dissipative effects) the *mean* power input must come out to be zero. Taking the equations already developed, and assuming the steady-state solution, we have

$$F = F_0 \cos \omega t$$

$$x = \frac{F_0/m}{\omega_0^2 - \omega^2} \cos \omega t = C \cos \omega t$$

Therefore,

$$v = -\omega C \sin \omega t$$

$$P = -\omega C F_0 \sin \omega t \cos \omega t$$

This power input, being proportional to $\sin 2\omega t$, is positive half the time and negative for the other half, averaging out to zero over any integral number of half-periods of oscillation. That is, energy is fed into the system during one quarter-cycle and is taken out again during the next quarter-cycle.

Coming now to the forced oscillator with damping, we have

$$x = A \cos(\omega t - \delta)$$

Therefore,

$$v = -\omega A \sin(\omega t - \delta)$$

We can write this as

$$v = -v_0 \sin(\omega t - \delta)$$

where v_0 is the maximum value of v for any given values of F_0 and ω. Taking the value of A from Eq. (4-14) we have

$$v_0(\omega) = \frac{F_0\omega_0/k}{\left[\left(\frac{\omega_0}{\omega} - \frac{\omega}{\omega_0}\right)^2 + \frac{1}{Q^2}\right]^{1/2}} \tag{4-21}$$

The value of v_0 passes through a maximum at $\omega = \omega_0$, exactly, a phenomenon that we can call velocity resonance.

Now let us consider the work and the power needed to maintain the forced oscillations. We have

$$P = -F_0 v_0 \cos \omega t \sin(\omega t - \delta)$$

$$= -F_0 v_0 \cos \omega t (\sin \omega t \cos \delta - \cos \omega t \sin \delta)$$

i.e.,

$$P = -(F_0v_0 \cos \delta) \sin \omega t \cos \omega t + (F_0v_0 \sin \delta) \cos^2 \omega t \quad (4\text{--}22)$$

If we average the power input over any integral number of cycles the first term in Eq. (4–22) gives zero. The average of $\cos^2 \omega t$, however, is $\frac{1}{2}$,[1] so that the average power input is given by

$$\bar{P} = \tfrac{1}{2}F_0v_0 \sin \delta = \tfrac{1}{2}\omega AF_0 \sin \delta$$

With the help of Eqs. (4–14) and (4–21) this becomes

$$\bar{P}(\omega) = \frac{F_0^2\omega_0}{2kQ} \, \frac{1}{\left(\dfrac{\omega_0}{\omega} - \dfrac{\omega}{\omega_0}\right)^2 + \dfrac{1}{Q^2}} \quad (4\text{--}23)$$

We see that this power input, like the velocity, passes through a maximum at *precisely* $\omega = \omega_0$ for any Q. The maximum power is given by

$$P_m = \frac{F_0^2\omega_0 Q}{2k} = \frac{QF_0^2}{2m\omega_0} \quad (4\text{--}24)$$

The dependence of $\bar{P}$ on ω for various Q is shown in Fig. 4–12(a). It may be noted that the power input drops off toward zero for very low and very high frequencies, and that except for low Q the curves are nearly symmetrical about the maximum. *It is convenient to define a width for these power resonance curves by taking the difference between those values of ω for which the power input is half of the maximum value.* This can be done in a particularly clear and useful way if (as in most cases of interest) Q is large. This means that the resonance is effectively contained within a narrow band of frequencies close to ω_0. It is then possible to write an approximate form of the equation for $\bar{P}(\omega)$, based on the following piece of algebra:

$$\frac{\omega_0}{\omega} - \frac{\omega}{\omega_0} = \frac{\omega_0^2 - \omega^2}{\omega\omega_0}$$

$$= \frac{(\omega_0 + \omega)(\omega_0 - \omega)}{\omega\omega_0}$$

Hence, if $\omega \approx \omega_0$, we can put

$$\frac{\omega_0}{\omega} - \frac{\omega}{\omega_0} \approx \frac{2\omega_0(\omega_0 - \omega)}{\omega_0^2} = \frac{2(\omega_0 - \omega)}{\omega_0}$$

Substituting this in the denominator of Eq. (4–23), we have

[1]Recall, for example, that $\cos^2 \omega t = \frac{1}{2}(1 + \cos 2\omega t)$ and that $(\cos 2\omega t)_{av} = 0$ over a complete cycle.

$$\overline{P}(\omega) = \frac{F_0{}^2 \omega_0}{2kQ} \frac{1}{\dfrac{4(\omega_0 - \omega)^2}{\omega_0{}^2} + \dfrac{1}{Q^2}}$$

$$= \frac{F_0{}^2 (\omega_0/Q)}{2(k/\omega_0{}^2)} \frac{1}{4(\omega_0 - \omega)^2 + (\omega_0/Q)^2}$$

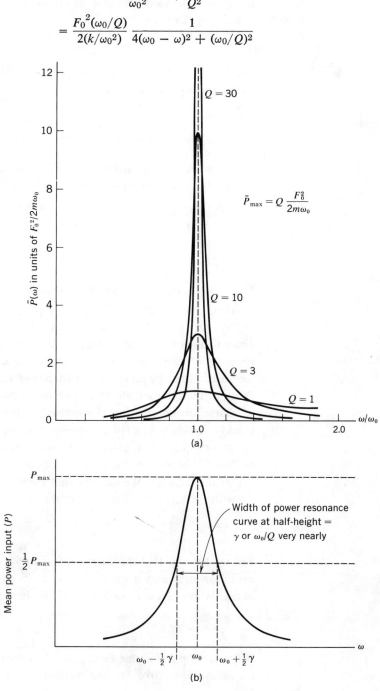

Fig. 4–12 (a) Mean power absorbed by a forced oscillator as a function of frequency for different values of Q. (b) Sharpness of resonance curve determined in terms of power curve.

Power absorbed by a driven oscillator

Now we have met the quantity ω_0/Q before. It is the damping constant $\gamma \ (= b/m)$ which characterizes the rate at which the energy of a damped oscillator was found to decay in the absence of a driving force:

$$E = E_0 e^{-(\omega_0/Q)t} = E_0 e^{-\gamma t} \tag{4-25}$$

[see Eq. (3–36)]. Thus the above equation for $\bar{P}$ can be written (remembering also that $k = m\omega_0{}^2$) in the following simplified form:

$$(\text{approximate}) \quad \bar{P}(\omega) = \frac{\gamma F_0{}^2}{2m} \ \frac{1}{4(\omega_0 - \omega)^2 + \gamma^2} \tag{4-26}$$

The frequencies $\omega_0 \pm \Delta\omega$ at which $\bar{P}(\omega)$ falls to half of the maximum value $\bar{P}(\omega_0)$ are thus defined by

$$4(\Delta\omega)^2 = \gamma^2$$

i.e.,

$$2\,\Delta\omega \approx \frac{\omega_0}{Q} \tag{4-27}$$

Thus we find that the width of the resonance curve for the *driven* oscillator, as measured by the power input [Fig. 4–12(b)], is equal to the reciprocal of the time needed for the *free* oscillations to decay to $1/e$ of their initial energy. We can thus predict that if a system is observed to have a very narrow resonance response (as measured either by amplitude or by power absorption), then the decay of its free oscillations will be very slow. And conversely, of course, an observation of whether the free oscillations decay quickly or slowly will tell us whether the response of the driven oscillator is broad or narrow. What is our criterion of "slow" or "fast," "broad" or "narrow"? Equations (4–26) and (4–27) tell us the answer. We can say that the resonance is narrow if the width is only a small fraction of the resonant frequency, i.e., if

$$\frac{2\,\Delta\omega}{\omega_0} \ll 1 \tag{4-28a}$$

and we can say that the decay of free oscillations is slow if the oscillator loses only a small fraction of its energy in one period of oscillation. Now from Eq. (4–25) we have

$$\frac{\Delta E}{E} \approx -\gamma\,\Delta t$$

If for Δt we put the time $2\pi/\omega_0$, which is approximately equal to the period of the free damped oscillation [Eq. (3–40)], we have

$$\frac{\Delta E}{E} \approx - \frac{2\pi\gamma}{\omega_0}$$

Thus a slow decay means

$$\frac{2\pi\gamma}{\omega_0} \ll 1 \tag{4-28b}$$

Since $\gamma = 2\,\Delta\omega = \omega_0/Q$, the conditions described by Eqs. (4–28a) and (4–28b) can both be expressed by saying that the dimensionless quantity Q must be large.

This relation between the resonance width of forced oscillation and the decrement of free oscillations is characteristic of a wide variety of oscillatory physical systems, not only the mechanical oscillator which we are here using as an example. In fact, whenever such a physical system, in free oscillation, shows an exponential loss of energy with time, it also displays a driven response having resonance characteristics.

EXAMPLES OF RESONANCE

In the course of our discussions we have made passing references to the fact that many systems which, on the face of it, have very little in common with a mass on a spring, nevertheless exhibit a similar resonance behavior. In concentrating on the behavior of a simple mechanical system, however, our analysis became very detailed and specific. Now we shall broaden our view again, and say something about resonance in quite different systems.

If we are to extend our ideas in this way, we need to be able to say in rather general terms what we mean by resonance, and we can begin by asking ourselves: What is the real essence of the behavior of the mass and spring system? And putting aside the mathematics we can say this: The system is acted on by an external agency, one parameter of which (the frequency) is varied. The response of the system, as measured by its amplitude and phase, or by the power absorbed, undergoes rapid changes as the frequency passes through a certain value. The form of the response is described by two quantities—a frequency ω_0 and a width γ $(= \omega_0/Q)$—which characterize the distinctive properties of the driven system. Resonance is the phenomenon of driving the system under such conditions that the interaction between the driving agency and the system is maximized. Whatever the particular criteria applied, one can say that the interaction has its maximum at or near ω_0, and that its most marked changes occur over a range of about $\pm\gamma$ with respect to the maximum.

When we carry over these ideas to the resonance behavior of other physical systems, we shall find that the quantities that characterize a resonance are not always frequency, absorbed power, and amplitude. This will appear in some of the examples that we shall now discuss.

ELECTRICAL RESONANCE

One of the most familiar and important resonant systems is the electrical system made up of a capacitor and a coil, as shown in Fig. 4–13. The analysis of such a system has a remarkable similarity to the mechanical systems with which we have been concerned so far. Let us consider first the free oscillations, ignoring for the moment any dissipative process associated with the electrical resistance. To begin with, we shall briefly describe the essential electrical behavior of the individual components.

The capacitor is a device for storing electric charge and the associated electrostatic potential energy. Its capacitance C is defined as the measure of the charge q applied to the capacitor plates divided by the measure of the voltage difference that this charge produces:

$$C = \frac{q}{V_C}$$

Therefore,

$$V_C = \frac{q}{C}$$

The action of the coil requires a somewhat more detailed description. Under D-C conditions the coil offers no opposition to the flow of current, but if the current is changing with time it is found that the coil (which we shall henceforth call an inductor) acts to oppose that change (Lenz's law). Under these circumstances

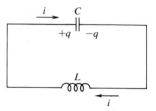

Fig. 4–13 Capacitor and inductor in series: the basic electrical resonance system.

there is a voltage difference V_L between the ends of the inductor, and this voltage is proportional to the rate of change of the current i. The inductance L is defined by the relation

$$V_L = L\frac{di}{dt}$$

This equation says that a voltage V_L must be applied between the ends of the inductor in order to make the current change at the rate di/dt.

In a circuit made up of just these two components, the sum of V_C and V_L must be zero, because an imaginary journey through the capacitor and then through the inductor brings us back to the same point on the circuit. Thus we have

$$\frac{q}{C} + L\frac{di}{dt} = 0 \tag{4-29}$$

Now there is an intimate connection between q and i, because the current in the circuit is just the rate of flow of charge past any point. A current i flowing for a time dt in the wire connected to a capacitor plate will increase the charge on that plate by the amount $dq = i\,dt$, so we have

$$i = \frac{dq}{dt}$$

$$\frac{di}{dt} = \frac{d^2q}{dt^2}$$

Hence Eq. (4-29) can be written

$$L\frac{d^2q}{dt^2} + \frac{1}{C}q = 0 \tag{4-30}$$

But this is precisely like the basic differential equation of SHM for a mass–spring system, with q playing the role of x, L appearing in the place of m, and $1/C$ replacing the spring constant k. We can confidently assume the existence of free electrical oscillations such that

$$\omega_0 = \frac{1}{\sqrt{LC}}$$

Now let us consider the effect of introducing a resistor, of resistance R, as in Fig. 4–14(a). At current i it is necessary to have a voltage V_R ($= iR$) applied between the ends of the resistor. Thus the statement of zero net voltage drop in one complete tour of the circuit is as follows:

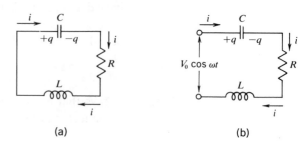

Fig. 4-14 (a) Capacitor, inductor, and resistor in series. (b) Capacitor, inductor, and resistor in series driven by a sinusoidal voltage.

(a)

(b)

$$\frac{q}{C} + iR + L\frac{di}{dt} = 0$$

i.e.,

$$L\frac{d^2q}{dt^2} + R\frac{dq}{dt} + \frac{1}{C}q = 0$$

or

$$\frac{d^2q}{dt^2} + \frac{R}{L}\frac{dq}{dt} + \frac{1}{LC}q = 0 \qquad (4\text{-}31)$$

In this equation, R/L plays exactly the role of the damping constant γ, and in such a circuit the charge on the capacitor plates (and the voltage V_C) will undergo exponentially damped harmonic oscillations.

Finally, if the circuit is driven by an alternating applied voltage, we have a typical forced-oscillator equation:

$$\frac{d^2q}{dt^2} + \frac{R}{L}\frac{dq}{dt} + \frac{1}{LC}q = \frac{V_0}{L}\cos\omega t \qquad (4\text{-}32)$$

Compare:

$$\frac{d^2x}{dt^2} + \frac{b}{m}\frac{dx}{dt} + \frac{k}{m}x = \frac{F_0}{m}\cos\omega t \qquad (4\text{-}33)$$

The connection between Eqs. (4-32) and (4-33) becomes even closer if one considers the energy of the system. Just as $F\,dx$ is the amount of work done by the driving force F in a displacement dx, so $V\,dq$ is the amount of work done by the driving voltage V when an amount of charge dq passes through the circuit. One can regard the oscillation as involving the periodic transfer of energy between the capacitor and the inductor, with a continual dissipation of energy in the resistor. Comparison of the mechanical and electrical equations suggests the classification of analogous quantities, as shown in Table 4-2.

We have discussed this phenomenon of electrical resonance

TABLE 4–2: MECHANICAL AND ELECTRICAL RESONANCE
PARAMETERS

Mechanical system	Electrical system
Displacement x	Charge q
Driving force F	Driving voltage V
Mass m	Inductance L
Viscous force constant b	Resistance R
Spring constant k	Reciprocal capacitance $1/C$
Resonant frequency $\sqrt{k/m}$	Resonant frequency $1/\sqrt{LC}$
Resonance width $\gamma = b/m$	Resonance width $\gamma = R/L$
Potential energy $\frac{1}{2}kx^2$	Energy of static charge $\frac{1}{2}q^2/C$
Kinetic energy $\frac{1}{2}m(dx/dt)^2 = \frac{1}{2}mv^2$	Electromagnetic energy of moving charge $\frac{1}{2}L(dq/dt)^2 = \frac{1}{2}Li^2$
Power absorbed at resonance $F_0{}^2/2b$	Power absorbed at resonance $V_0{}^2/2R$

at some length because of its extremely close likeness to mechanical resonance. Our other examples, although of great physical importance, do not fall so completely into this pattern, and we shall dispose of them more briefly.

OPTICAL RESONANCE

We have a great wealth of evidence that atoms behave like sharply tuned oscillators in the processes of emitting and absorbing light. Whenever the emission of light occurs under such conditions that the radiating atoms are effectively isolated from each other, as in a gas at low pressure, the spectrum consists of discrete, very narrow lines; i.e., the radiated energy is concentrated at particular wavelengths. An incandescent solid—e.g., the filament of a light bulb—emits a continuous spectrum, but the situation here is quite different, because each atom in a solid is strongly linked to its neighbors, causing a drastic change in the dynamical state of the electrons chiefly responsible for visible or near-visible radiation.

We have just spoken of atoms as oscillators that emit their characteristic frequencies. But how does this fit in with the photon description of radiation, and with the picture of the radiative process as one in which the atom undergoes a quantum jump? The answer is by no means obvious. Before the advent of quantum theory, one could visualize an electron describing a circular orbit within an atom, and emitting light of a frequency equal to its own orbital frequency. But now we can only say that the frequency of the light is defined (through $E = h\nu$) by the energy difference between two states of the atom; we can no longer identify that frequency with a vibration of the atom itself. Nevertheless the concept of the atom as an oscillator does in some respects survive. If the emitted light is analyzed with an interferometer, it is found to consist of wave trains of finite length. The length of the wave trains, divided by c, defines a time τ which corresponds to the mean life of the radiating atoms in their excited state, and the surplus energy of a collection of excited atoms decays exponentially as $e^{-t/\tau}$ $(= e^{-\gamma t})$ as the energy is radiated away. Neither the photon picture nor the wave picture alone tells us the whole story, but the model of the atom as a damped oscillator provides an acceptable description of some important aspects of the radiative process.

As we have seen, the concomitant of a natural frequency of free oscillation is a resonance absorption at about that same frequency. In the case of visible light the frequencies are too high ($\approx 10^{15}$ Hz) to be measurable, but we are able to describe both emission and absorption in terms of characteristic wavelengths. Probably the most famous example of resonance absorption for light is provided by the Fraunhofer lines. These are the dark lines that are observed in a spectrum analysis of the sun; they are named after Joseph von Fraunhofer, who in a careful study mapped 576 of them in 1814. Figure 4–15(a) shows a portion of

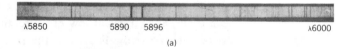

(a)

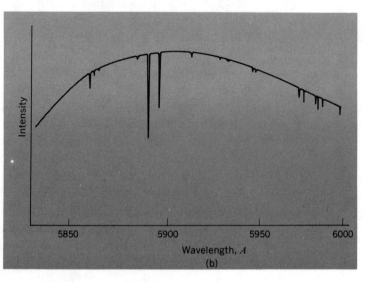

Fig. 4–15 (a) Portion of the solar spectrum, showing the famous sodium D lines at 5890 and 5896 Å. (From F. A. Jenkins and H. E. White, Fundamentals of Optics, *McGraw-Hill, New York, 1957.) (b) A qualitative representation of the intensity of the solar spectrum as a function of wavelength, over the range shown in (a).*

Forced vibrations and resonance

the solar spectrum; the prominent Fraunhofer lines at 5890 and 5896 Å are due to sodium. Figure 4–15(b) shows qualitatively what a plot of intensity versus wavelength looks like; the intensity dips sharply at the wavelength of the Fraunhofer lines, but is not zero. (It was not Fraunhofer who first observed the absorption lines,[1] but it was he who first recognized that some of them coincided in wavelength with bright emission lines produced by laboratory sources. It remained, however, for Kirchhoff and Bunsen in 1861 to make a detailed comparison of the solar spectrum with the arc and spark spectra of pure elements.)

One can be sure that the Fraunhofer lines are the result of resonance absorption processes. The picture is that the continuous radiation from hot and relatively dense matter near the sun's surface is selectively filtered, as it passes outward, by atoms in the more tenuous vapors of the solar atmosphere. It would be satisfying if one could trace out the detailed shape of an optical absorption line and relate its width to the characteristic time ($= 1/\gamma$) for the decay of the spontaneous emission. This, however, is extremely hard to do. The chief enemy is the Doppler effect. Both direct and indirect evidence show that a typical lifetime for an excited atom emitting visible light is about 10^{-8} sec, so that γ is about 10^8 sec^{-1}. The angular frequency of the emitted light, as defined by $2\pi c/\lambda$, is about 4×10^{15} sec^{-1}. Thus we can calculate a line width $\delta\lambda$ as follows:

$$\frac{\delta\lambda}{\lambda} \approx \frac{\delta\omega}{\omega_0} = \frac{\gamma}{\omega_0} \approx \frac{10^8}{4 \times 10^{15}} \approx 2 \times 10^{-8}$$

(Hence $\delta\lambda \approx 10^{-4}$Å for $\lambda \approx 5000$ Å.) But, unless special precautions are taken, the emitting atoms have random thermal motions of several hundred meters per second, and we can estimate a Doppler broadening of the spectral lines:

$$\frac{\Delta\lambda}{\lambda} = \frac{v}{c} \approx 10^{-6}$$

The Doppler effect is thus about 100 times greater than any effect due to the true lifetime of the radiating atom. Interatomic collisions also disturb the situation, so that the resonance shapes of spectral lines are more a matter of inference than of direct spectroscopic observation.

[1]They were first noted by W. H. Wollaston in 1802. By 1895 a classic study by the American physicist H. A. Rowland had resulted in the mapping of 1100 of them. Today about 26,000 lines have been catalogued between 3000 and 13,000 Å.

107   Optical resonance

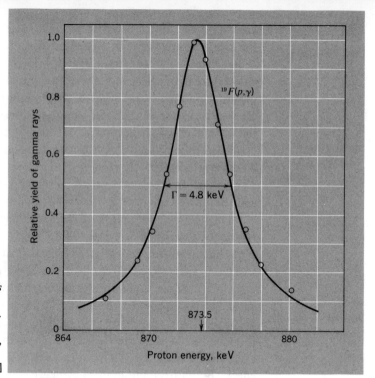

Fig. 4-16   Yield of gamma rays as a function of the energy of bombarding protons in the reaction $p + {}^{19}F \rightarrow {}^{20}Ne + \gamma$. [From data of R. G. Herb, S. C. Snowden, and O. Sala, Phys. Rev., **75**, 246 (1949).]

NUCLEAR RESONANCE

The literature of nuclear physics contains innumerable examples of nuclear resonances; Fig. 4–16 shows one of them. This process of nuclear resonance differs in several ways from anything we have discussed so far. The subject of Fig. 4–16 is a nuclear reaction; the graph shows the relative yield of gamma rays as a target of fluorine is bombarded with protons of different energies around 875 keV. But what is the resonant system? It is not the bombarded fluorine but the compound nucleus—${}^{20}Ne$ in an excited state, denoted ${}^{20}Ne^*$, formed when a fluorine nucleus captures a proton. This compound nucleus is unstable, and one of its decay modes is by emission of gamma rays. The complete process can be written as follows:

$$\mathrm{{}^{1}_{1}H} + \mathrm{{}^{19}_{9}F} \rightarrow \mathrm{{}^{20}_{10}Ne^*} \rightarrow \mathrm{{}^{20}_{10}Ne} + \gamma$$

(The subscript shows the number of protons in a nucleus, and the superscript the total of protons plus neutrons.)

The controllable parameter—the independent variable of the interaction—is not a frequency but the energy of the bombarding proton. This defines a basic property of the resonance: the total energy of the ^{20}Ne* in its rest frame. The response of the system is measured, not in terms of amplitude or absorbed power, but in terms of the probability that an incident proton will cause a gamma ray to be produced. This probability can be described in terms of the effective target area (or cross section, σ) that each fluorine nucleus presents to the incident proton beam. Finally, the detailed shape of the resonance curve is very similar in analytic form to the approximate form (for high Q) of the absorbed power curve of a mechanical oscillator [Eq. (4–26) and Fig. 4–12]. A nuclear resonance such as the one of Fig. 4–16 can be well described by the equation

$$\sigma(E) = \frac{\sigma(E_0)}{\dfrac{4(E_0 - E)^2}{\Gamma^2} + 1} \tag{4–34}$$

The energy E_0 then corresponds to the peak of the resonance curve, and the total width of the curve at half-height is given by Γ. Defined in this way, the energy width Γ is strictly analogous to the frequency width γ of a mechanical or electrical resonance. The full curve in Fig. 4–16 is drawn according to Eq. (4–34) with appropriate values of E_0 and Γ, and it can be seen that the fit to the data is excellent.

NUCLEAR MAGNETIC RESONANCE

As a last example of resonance in other fields of physics, we shall mention the resonant process by which atomic nuclei, behaving as tiny magnets, can be flipped over in a magnetic field. It depends upon a quantum phenomenon: that atomic magnets are limited to having only a few discrete possible orientations with respect to a magnetic field in a given direction. A proton, to take a specific example, has only two possible orientations, one corresponding roughly to the north-seeking orientation of an ordinary compass needle, and the other corresponding to the reverse of this. There is a well-defined energy difference between these orientations, corresponding to the work done against the magnetic forces in turning the nuclear magnet from one position to the other. This energy difference is directly proportional to the strength of the magnetic field in which the proton finds itself. If photons of just the right energy come along, they can cause the protons to switch from one orientation to the other. This can be brought about by injecting electromagnetic radiation of just the right frequency; for protons in a field of about 5000 G the resonance frequency is about 21 MHz. If all the protons in about 1 cm^3 of water are flipped in this way, they can be made to produce (through electromagnetic induction) a readily detectable voltage in a pickup coil. If the magnetic field were held constant, one would see this signal as a resonant function of the frequency of the injected radiation. It is much more convenient, however, to use a constant, sharply defined radiofrequency and vary the strength of the applied magnetic field B. The magnitude of the nuclear magnetic resonance signal can then be expressed as a resonant function of the field strength:

$$V(B) = \frac{V_0}{\dfrac{4(B_0 - B)^2}{(\Delta B)^2} + 1} \qquad (4\text{-}35)$$

where B_0 is the field strength at exact resonance and ΔB is the width of the resonance at half-height.

For their quite independent research on this phenomenon,

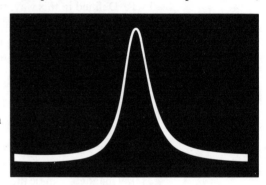

Fig. 4-17   Magnetic resonance line of protons in water containing $MnSO_4$ as a paramagnetic catalyst and obtained from that component of the nuclear induction signal which corresponds to absorption. The photograph is of the trace on a cathode-ray oscillograph with the vertical deflection arising from the rectified and amplified signal and the horizontal deflection corresponding to different values of the constant field.   From Nobel Lectures: Physics (1942–1962), Elsevier, Amsterdam, 1964.

F. Bloch and E. M. Purcell shared the Nobel Prize in physics in 1952. Figure 4-17 comes from the Nobel lecture that Bloch gave at that time.

ANHARMONIC OSCILLATORS

So far this chapter reads altogether too much like a success story. Everything works. We write down a differential equation and obtain in every case an analytic solution that fits it exactly. We point to actual physical systems that apparently conform perfectly to our very simple mathematical model. Is nature really so accommodating? The answer is that in certain cases—numerous and varied enough to be of great physical importance—a system can indeed be represented, with impressive accuracy, as a damped oscillator with a restoring force proportional to the displacement and a resistive force proportional to the velocity. But this is an astonishing stroke of luck, and we have in fact been treading a very narrow path. To appreciate just how special and favorable are the situations that we have discussed, we shall glance briefly at the effect of modifying the equations of motion.

Our original equation for the free oscillation of a mass on a spring without damping was the following:

$$F = m\frac{d^2x}{dt^2} = -kx$$

This holds if the spring obeys a linear relation (Hooke's law) for any amount of extension or compression. But no real spring behaves quite like this. With many springs it takes a slightly different size of force to produce a given extension than to produce an equal compression. The simplest asymmetry of this kind is represented by a term in F proportional to x^2. Or it may be that the spring is symmetrical with respect to positive and negative displacements, but that there is not strict proportionality of F to x. The simplest symmetrical effect of this kind is described by a term in F proportional to x^3. The equations of motion for these cases can be written as follows:

$$\text{Nonlinear, asymmetric: } m\frac{d^2x}{dt^2} + kx + \alpha x^2 = 0 \qquad (4\text{--}36a)$$

$$\text{Nonlinear, symmetric: } m\frac{d^2x}{dt^2} + kx + \beta x^3 = 0 \qquad (4\text{--}36b)$$

If we try a solution of the form $x = A \cos \omega_0 t$ in either of the above equations we find at once that it does not work; the motion is no longer describable as a harmonic vibration at some unique frequency ω_0. We have instead what is called an anharmonic oscillator. The motion is still periodic, in that (assuming no damping) a given state of the motion recurs at equal intervals $T = 2\pi/\omega_0$, but instead of having $x = A \cos \omega_0 t$ we find that an infinite set of harmonics of ω_0 is now needed to describe the motion; i.e., we must put

$$x = \sum_{n=1}^{\infty} A_n \cos(n\omega_0 t - \delta_n)$$

in order to have a form of x that will satisfy the differential equations.

In similar fashion, a resistive force varying as v^2 or v^3, instead of v, makes impossible a clean, simple analytic description of the motion of a damped oscillator.

What happens if an oscillator with nonlinear terms (in restoring force, damping force, or both) is subjected to a sinusoidal driving force? We shall not try to spell out the answer but leave it as a challenge for your spare moments. Take, for example, an oscillator whose free oscillations are described by Eq. (4–36a) with a pure viscous force ($\sim dx/dt$) added, and assume a driving force $F = F_0 \cos \omega t$. Assume $\alpha x^2 \ll kx$, put $k/m = \omega_0{}^2$, and see if you can determine the frequency or frequencies ω for which the system exhibits resonance behavior. After investigating this problem you will realize that the simple harmonic oscillator is well named, and you will appreciate why a physicist will use it as a model of a vibratory system if it can possibly be justified.

PROBLEMS

4–1 Construct a table, covering as wide a range as possible, of resonant systems occurring in nature. Indicate the order of magnitude of (a) the physical size of each system, and (b) its resonant frequency.

4–2 Consider how to solve the steady-state motion of a forced oscillator if the driving force is of the form $F = F_0 \sin \omega t$ instead of $F_0 \cos \omega t$.

4–3 An object of mass 0.2 kg is hung from a spring whose spring constant is 80 N/m. The body is subject to a resistive force given by $-bv$, where v is its velocity (m/sec) and $b = 4$ N-m^{-1} sec.

(a) Set up the differential equation of motion for free oscillations of the system, and find the period of such oscillations.

(b) The object is subjected to a sinusoidal driving force given by $F(t) = F_0 \sin \omega t$, where $F_0 = 2$ N and $\omega = 30$ sec^{-1}. In the steady state, what is the amplitude of the forced oscillation?

4–4 A block of mass m is connected to a spring, the other end of which is fixed. There is also a viscous damping mechanism. The following observations have been made on this system:

(1) If the block is pushed horizontally with a force equal to mg, the static compression of the spring is equal to h.

(2) The viscous resistive force is equal to mg if the block moves with a certain known speed u.

(a) For this complete system (including both spring and damper) write the differential equation governing horizontal oscillations of the mass in terms of m, g, h, and u.

Answer the following for the case that $u = 3\sqrt{gh}$:

(b) What is the angular frequency of the damped oscillations?

(c) After what time, expressed as a multiple of $\sqrt{h/g}$, is the *energy* down by a factor $1/e$?

(d) What is the Q of this oscillator?

(e) This oscillator, initially in its rest position, is suddenly set into motion at $t = 0$ by a bullet of negligible mass but nonnegligible momentum traveling in the positive x direction. Find the value of the phase angle δ in the equation $x = Ae^{-\gamma t/2} \cos(\omega t - \delta)$ that describes the subsequent motion, and sketch x versus t for the first few cycles.

(f) If the oscillator is driven with a force $mg \cos \omega t$, where $\omega = \sqrt{2g/h}$, what is the amplitude of the steady-state response?

4-5 A simple pendulum has a length (l) of 1 m. In free vibration the amplitude of its swings falls off by a factor e in 50 swings. The pendulum is set into forced vibration by moving its point of suspension horizontally in SHM with an amplitude of 1 mm.

(a) Show that if the horizontal displacement of the pendulum bob is x, and the horizontal displacement of the support is ξ, the equation of motion of the bob for small oscillations is

$$\frac{d^2x}{dt^2} + \gamma \frac{dx}{dt} + \frac{g}{l} x = \frac{g}{l} \xi$$

Solve this equation for steady-state motion, if $\xi = \xi_0 \cos \omega t$. (Put $\omega_0^2 = g/l$.)

(b) At exact resonance, what is the amplitude of the motion of the pendulum bob? (First, use the given information to find Q.)

(c) At what angular frequencies is the amplitude half of its resonant value?

4-6 Imagine a simple seismograph consisting of a mass M hung from a spring on a rigid framework attached to the earth, as shown. The spring force and the damping force depend on the displacement and velocity relative to the earth's surface, but the dynamically significant acceleration is the acceleration of M relative to the fixed stars.

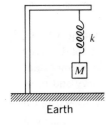

Earth

(a) Using y to denote the displacement of M relative to the earth and η to denote the displacement of the earth's surface itself, show that the equation of motion is

$$\frac{d^2y}{dt^2} + \gamma \frac{dy}{dt} + \omega_0^2 y = -\frac{d^2\eta}{dt^2}$$

(b) Solve for y (steady-state vibration) if $\eta = C \cos \omega t$.

(c) Sketch a graph of the amplitude A of the displacement y as a function of ω (supposing C the same for all ω).

(d) A typical long-period seismometer has a period of about 30 sec and a Q of about 2. As the result of a violent earthquake the earth's surface may oscillate with a period of about 20 min and with an amplitude such that the maximum acceleration is about 10^{-9} m/sec². How small a value of A must be observable if this is to be detected?

4-7 Consider a system with a damping force undergoing forced oscillations at an angular frequency ω.

(a) What is the instantaneous kinetic energy of the system?

(b) What is the instantaneous potential energy of the system?

(c) What is the ratio of the average kinetic energy to the average potential energy? Express the answer in terms of the ratio ω/ω_0.

(d) For what value(s) of ω are the average kinetic energy and the average potential energy equal? What is the total energy of the system under these conditions?

(e) How does the total energy of the system vary with time for an arbitrary value of ω? For what value(s) of ω is the total energy constant in time?

4–8 A mass m is subject to a resistive force $-bv$ but *no* springlike restoring force.

(a) Show that its displacement as a function of time is of the form

$$x = C - \frac{v_0}{\gamma} e^{-\gamma t}$$

where $\gamma = b/m$.

(b) At $t = 0$ the mass is at rest at $x = 0$. At this instant a driving force $F = F_0 \cos \omega t$ is switched on. Find the values of A and δ in the steady-state solution $x = A \cos(\omega t - \delta)$.

(c) Write down the general solution [the sum of parts (a) and (b)] and find the values of C and v_0 from the conditions that $x = 0$ and $dx/dt = 0$ at $t = 0$. Sketch x as a function of t.

4–9 (a) A forced damped oscillator of mass m has a displacement varying with time given by $x = A \sin \omega t$. The resistive force is $-bv$. From this information calculate how much work is done against the resistive force during one cycle of oscillation.

(b) For a driving frequency ω *less* than the natural frequency ω_0, sketch graphs of potential energy, kinetic energy, and total energy for the oscillator over one complete cycle. Be sure to label important turning points and intersections with their values of energy and time.

4–10 The power input to maintain forced vibrations can be calculated by recognizing that this power is the mean rate of doing work against the resistive force $-bv$.

(a) Satisfy yourself that the instantaneous rate of doing work against this force is equal to bv^2.

(b) Using $x = A \cos(\omega t - \delta)$, show that the mean rate of doing work is $b\omega^2 A^2/2$.

(c) Substitute the value of A at any arbitrary frequency and hence obtain the expression for $\bar{P}$ as given in Eq. (4–23).

4–11 Consider a damped oscillator with $m = 0.2$ kg, $b = 4$ N-m^{-1} sec and $k = 80$ N/m. Suppose that this oscillator is driven by a force $F = F_0 \cos \omega t$, where $F_0 = 2$ N and $\omega = 30$ sec^{-1}.

(a) What are the values A and δ of the steady-state response described by $x = A \cos(\omega t - \delta)$?

(b) How much energy is dissipated against the resistive force in one cycle?

(c) What is the mean power input?

4–12 An object of mass 2 kg hangs from a spring of negligible mass. The spring is extended by 2.5 cm when the object is attached. The top end of the spring is oscillated up and down in SHM with an amplitude of 1 mm. The Q of the system is 15.

(a) What is ω_0 for this system?

(b) What is the amplitude of forced oscillation at $\omega = \omega_0$?

(c) What is the mean power input to maintain the forced oscillation at a frequency 2% greater than ω_0? [Use of the approximate formula, Eq. (4–26), is justified.]

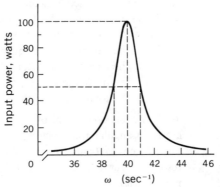

4–13 The graph shows the power resonance curve of a certain mechanical system when driven by a force $F_0 \sin \omega t$, where $F_0 =$ constant and ω is variable.

(a) Find the numerical values of ω_0 and Q for this system.

(b) The driving force is turned off. After how many cycles of free oscillation is the energy of the system down to $1/e^5$ of its initial value? ($e = 2.718$.) (To a good approximation, the period of free oscillation can be set equal to $2\pi/\omega_0$.)

4–14 The figure shows the mean power input $\overline{P}$ as a function of driving frequency for a mass on a spring with damping. (Driving force =

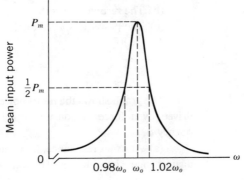

$F_0 \sin \omega t$, where F_0 is held constant and ω is varied.) The Q is high enough so that the mean power input, which is maximum at ω_0, falls to half-maximum at the frequencies $0.98\omega_0$ and $1.02\omega_0$.

(a) What is the numerical value of Q?

(b) If the driving force is removed, the energy decreases according to the equation

$$E = E_0 e^{-\gamma t}$$

What is the value of γ?

(c) If the driving force is removed, what fraction of the energy is lost per cycle?

A new system is made in which the spring constant is doubled, but the mass and viscous medium are unchanged, and the same driving force $F_0 \sin \omega t$ is applied. In terms of the corresponding quantities for the original system, find the values of the following:

(d) The new resonant frequency ω_0'.

(e) The new quality factor Q'.

(f) The maximum mean power input $\bar{P}_m'$.

(g) The total energy of the system at resonance, E_0'.

4–15 The free oscillations of a mechanical system are observed to have a certain angular frequency ω_1. The same system, when driven by a force $F_0 \cos \omega t$ (where F_0 = const. and ω is variable), has a power resonance curve whose angular frequency width, at half-maximum power, is $\omega_1/5$.

(a) At what angular frequency does the maximum power input occur?

(b) What is the Q of the system?

(c) The system consists of a mass m on a spring of spring constant k. In terms of m and k, what is the value of the constant b in the resistive term $-bv$?

(d) Sketch the amplitude response curve, marking a few characteristic points on the curve.

4–16 For the electrical system in the figure, find

(a) The resonant frequency, ω_0.

(b) The resonance width, γ.

(c) The power absorbed at resonance.

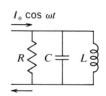

$I_0 \cos \omega t$

R C L

4–17 The graph shows the mean power absorbed by an oscillator when driven by a force of constant magnitude but variable angular frequency ω.

(a) At exact resonance, how much work per cycle is being done against the resistive force? (Period = $2\pi/\omega$.)

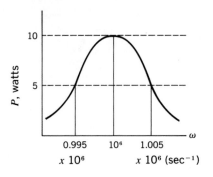

0.995 10^6 1.005

$x\ 10^6$ $x\ 10^6\ (\text{sec}^{-1})$

(b) At exact resonance, what is the *total* mechanical energy E_0 of the oscillator?

(c) If the driving force is turned off, how many seconds does it take before the energy decreases to a value $E = E_0 e^{-1}$?

The question of the vibration of connected particles is a peculiarly interesting and important problem . . . it is going to have many applications.

LORD KELVIN, *Baltimore Lectures* (1884)

5

Coupled
oscillators and
normal modes[1]

THROUGHOUT THE PRECEDING TWO CHAPTERS we have confined
our analysis to systems having only one type of free vibration,
and characterized by a single natural frequency. A real physical
system, however, is usually capable of vibrating in many different
ways, and may resonate to many different frequencies—like a
sort of grand piano. We speak of these various characteristic
vibrations as *modes*, or, for reasons that will emerge later, as
normal modes of the system. A simple example is a flexible chain
suspended from one end. It is found that there is a whole succession of frequencies at which every point on the chain vibrates
in SHM at the same frequency, so that the shape of the chain
remains constant in the sense that the displacements of the various
parts always preserve fixed ratios. The first three modes (in
ascending order of frequency) for such a chain are shown in
Fig. 5–1. This is in effect only a one-dimensional object, and the
variety of natural modes of oscillation for two- and three-
dimensional objects is still greater.

[1]This whole chapter may be bypassed if it is preferred to proceed directly to
the discussion of vibrations and waves in effectively continuous media. On
the other hand, an acquaintance with the contents of the present chapter, even
in rather general terms, may help in appreciating the sequel, for the many-
particle system does provide the natural link between the single oscillator
and the continuum. And it is not as mathematically formidable as it may
appear at first sight.

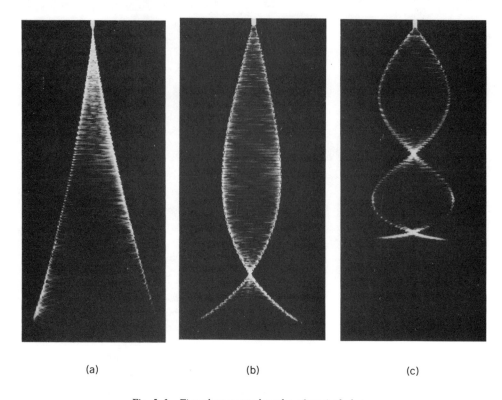

(a)   (b)   (c)

Fig. 5–1 First three normal modes of vertical chain with upper end fixed. (The tension is provided at each point by the weight of the chain below that point and so increases linearly with distance from the bottom.)

How do we go about the job of accounting for these numerous modes and calculating their frequencies? The clue to this question lies in the fact that an extended object can be regarded as a large number of simple oscillators coupled together. A solid body, for example, is composed of many atoms or molecules. Every atom may behave as an oscillator, vibrating about an equilibrium position. But the motion of each atom affects its neighbors so that, in effect, all the atoms of the solid are coupled together. The question then becomes: How does the coupling affect the behavior of the individual oscillators?

We shall begin by discussing in some detail the properties of a system of just two coupled oscillators. The change from one oscillator to two may seem rather trivial, but this new system has some novel and surprising features. Moreover, in analyzing its behavior we shall develop essentially all the theoretical tools we need to handle the problem of an arbitrarily large number of coupled oscillators—which will be our ultimate concern. And this means that, from quite simple beginnings, we can end up with a significant insight into the dynamical properties of something as complicated as a crystal lattice. That is no small achievement, and it is worth the little extra amount of mathematical effort that our discussion will entail.

TWO COUPLED PENDULUMS

Let us begin with a very simple example. Take two identical pendulums, *A* and *B*, and connect them with a spring whose relaxed length is exactly equal to the distance between the pendulum bobs, as shown in Fig. 5–2. Draw pendulum *A* aside while holding *B* fixed and then release both of them. What happens?

Pendulum *A* swings from side to side, but its amplitude of oscillation continuously decreases. Pendulum *B*, initially undisplaced, gradually begins to oscillate and its amplitude continuously increases. Soon, *A* and *B* have equal amplitudes. You might think that now there would be no further change. But no, the process continues. The amplitude of *A* continues to decrease and that of *B* to increase until eventually the displacement of *B* is equal (or about equal) to that originally given to *A*, and the displacement of *A* diminishes toward zero. The starting condition is almost reversed. Now it is easy to predict the sequel. The motion of *B* is transferred back to *A*, and so it continues. The energy, originally given to *A* (and to the spring), does not remain confined to the oscillation of *A*, but is transferred gradually to *B* and continues to shuttle back and forth between *A* and *B*. Figure 5–3 shows records of actual motions of such a coupled system. The pendulums, whose bobs were dry cells with flashlight bulbs attached, were suspended from the ceiling and were photographed from below by a camera that was pulled steadily along the floor.

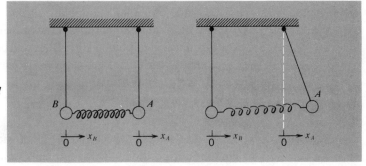

Fig. 5–2 (a) Coupled pendulums in equilibrium position. (b) Coupled pendulums with one pendulum displaced.

121   Two coupled pendulums

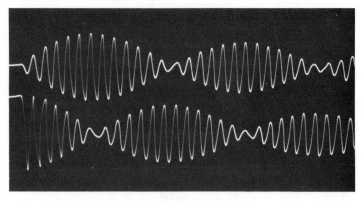

Fig. 5–3 Motion of two identical coupled oscillators (pendulums with flashlight bulbs on the bobs). Pendulum no. 1 was initially at rest at its normal equilibrium position. The damping of the system is quite noticeable. (Photo by Jon Rosenfeld, Education Research Center, M.I.T.)

Of course, it is the coupling spring that is responsible for the observed behavior. As A oscillates, the spring pulls and pushes on B. It provides a driving force that works on B and sets it into motion. At the same time, the spring pulls and pushes on A, sometimes helping, sometimes hindering its motion. But as B begins to move, the action of the spring on A is more to hinder than to help. The net work done on A during one oscillation is negative, and the amplitude of A decreases.

Each of the motions recorded in Fig. 5–3 looks just like a case of beats between two SHM's of the same amplitude but different frequencies. And that is precisely what they are. To account for them in detail is not, however, an obvious matter: Our "feeling" for the physical phenomenon helps us here only qualitatively. But the problem becomes exceedingly simple if we alter the starting conditions somewhat.

SYMMETRY CONSIDERATIONS

Suppose we draw both A and B aside by equal amounts [Fig. 5–4(a)] and then release them. The distance between them equals the relaxed length of the coupling spring and therefore the spring exerts no force on either pendulum. A and B will oscillate in phase and with equal amplitudes, always maintaining the same separation. Each pendulum might just as well be free (uncoupled). Each oscillates with its free natural frequency $\omega_0 \ (= \sqrt{g/l})$. The equations of motion are

$$x_A = C \cos \omega_0 t$$
$$x_B = C \cos \omega_0 t \tag{5-1}$$

where x_A and x_B are the displacements of each pendulum from its equilibrium position. This represents a *normal mode* of the

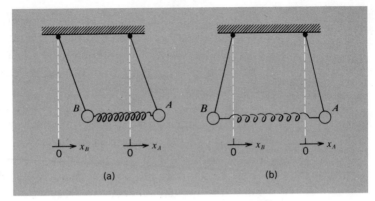

Fig. 5–4 (a) Lower normal mode of two coupled pendulums. (b) Higher normal mode of two coupled pendulums.

(a)   (b)

coupled system. Both masses vibrate at the same frequency and each has a constant amplitude (the same for both).

How many normal modes can we find? There is only one other. Draw A and B aside by equal amounts but in opposite directions [Fig. 5–4(b)] and then release them. Now, the coupling spring is stretched; a half-cycle later it will be compressed, and it does exert forces. The symmetry of the arrangement tells us that the motions of A and B will be mirror images of each other.

If the pendulums were free and either one were displaced a small distance x, the restoring force would be $m\omega_0^2 x$. But in the present situation the coupling spring is stretched (or compressed) a distance $2x$ and exerts a restoring force of $2kx$, where k is the spring constant. Thus the equation of motion for A is

$$m\frac{d^2 x_A}{dt^2} + m\omega_0^2 x_A + 2kx_A = 0$$

or

$$\frac{d^2 x_A}{dt^2} + (\omega_0^2 + 2\omega_c^2)x_A = 0$$

where we have let $\omega_c^2 = k/m$. This is an equation for simple harmonic motion of frequency ω' given by

$$\omega' = (\omega_0^2 + 2\omega_c^2)^{1/2} = \left(\frac{g}{l} + \frac{2k}{m}\right)^{1/2}$$

For the given starting conditions, its solution is

$$x_A = D \cos \omega't \tag{5–2a}$$

The motion of B is the mirror image of A, and therefore

$$x_B = -D \cos \omega't \tag{5–2b}$$

Each pendulum oscillates with simple harmonic motion, but the action of the coupling spring has been to increase the restoring force and therefore to increase the frequency over that of the uncoupled oscillation. The motions of A and B are clearly always 180° out of phase in this type of oscillation, which constitutes the second normal mode.

It is perhaps worth pointing out that if either of the pendulums is clamped, the angular frequency of the other, under the action of the gravity plus the coupling spring, is equal to $(\omega_0{}^2 + \omega_c{}^2)^{1/2}$. Thus if one chooses to regard this motion as being, in a sense, the motion characteristic of one pendulum alone, the normal modes have frequencies that are displaced above or below the single-pendulum value.

THE SUPERPOSITION OF THE NORMAL MODES

In both the above cases, the motion once begun will, in the absence of damping forces, continue without change. No transfer of energy occurs from some one mode of oscillation to another. An important reason for introducing these two easily solved cases is that any motion of the pendulums, in which each starts from rest, can be described as a combination of these two. Let us see how that can be done.

Take an arbitrary moment when pendulum A is at x_A and pendulum B at x_B (Fig. 5-5). The spring is stretched an amount $x_A - x_B$ and therefore pulls on A and B with a force whose magnitude is $k(x_A - x_B)$. Thus the magnitude of the restoring force on A is

$$m\omega_0{}^2 x_A + k(x_A - x_B)$$

and on B it is

$$m\omega_0{}^2 x_B - k(x_A - x_B)$$

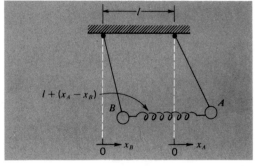

Fig. 5-5   Coupled
pendulums in arbitrary
configuration.

Therefore, the equations of motion for A and B are

$$m\frac{d^2x_A}{dt^2} + m\omega_0{}^2x_A + k(x_A - x_B) = 0$$

$$m\frac{d^2x_B}{dt^2} + m\omega_0{}^2x_B - k(x_A - x_B) = 0$$

(5-3)

Again letting $\omega_c{}^2 = k/m$, we can write these as follows:

$$\frac{d^2x_A}{dt^2} + (\omega_0{}^2 + \omega_c{}^2)x_A - \omega_c{}^2x_B = 0$$

$$\frac{d^2x_B}{dt^2} + (\omega_0{}^2 + \omega_c{}^2)x_B - \omega_c{}^2x_A = 0$$

(5-4)

The first equation, describing the acceleration of A, contains a term in x_B. And the second equation contains a term in x_A. These two differential equations cannot be solved independently but must be solved simultaneously. A motion given to A does not stay confined to A but affects B, and vice versa.

Actually, these equations are not difficult to solve. If we add the two together, we get

$$\frac{d^2}{dt^2}(x_A + x_B) + \omega_0{}^2(x_A + x_B) = 0$$

and if we subtract the second equation from the first, we get

$$\frac{d^2}{dt^2}(x_A - x_B) + (\omega_0 + 2\omega_c{}^2)(x_A - x_B) = 0$$

These are familiar equations for simple harmonic oscillations. In the first, the variable is $x_A + x_B$ and the frequency is ω_0. In the second, the variable is $x_A - x_B$ and the frequency is $\omega' = (\omega_0{}^2 + 2\omega_c{}^2)^{1/2}$. These two frequencies correspond precisely to those of the two normal modes that we identified previously. If we let $x_A + x_B = q_1$ and $x_A - x_B = q_2$, we have two independent equations in q_1 and q_2:

$$\frac{d^2q_1}{dt^2} + \omega_0{}^2q_1 = 0$$

$$\frac{d^2q_2}{dt^2} + \omega'^2q_2 = 0$$

Possible solutions (although not the most general ones) are

(special case) $\quad\begin{aligned}q_1 &= C\cos\omega_0 t\\ q_2 &= D\cos\omega' t\end{aligned}$

(5-5)

where C and D are constants which depend upon the initial conditions. [The lack of generality in Eqs. (5-5) can be recognized in the fact that we have set the initial phases equal to zero.]

We have here two independent oscillations. They represent another description of the normal modes, as represented by oscillations of the variables q_1 and q_2 respectively, and these variables are consequently called *normal coordinates*. Changes in the value of q_1 occur independently of q_2 and vice versa.

In terms of our original coordinates, x_A and x_B, the solutions are

$$\text{(special case)} \quad \begin{aligned} x_A &= \tfrac{1}{2}(q_1 + q_2) = \tfrac{1}{2}C \cos \omega_0 t + \tfrac{1}{2}D \cos \omega' t \\ x_B &= \tfrac{1}{2}(q_1 - q_2) = \tfrac{1}{2}C \cos \omega_0 t - \tfrac{1}{2}D \cos \omega' t \end{aligned} \quad (5\text{-}6)$$

If $C = 0$, both pendulums oscillate with the frequency ω', or if $D = 0$, with the frequency ω_0. These are the frequencies of the individual normal modes and are called *normal frequencies*. We see that a characteristic of a normal frequency is that both x_A and x_B can oscillate with that frequency.

Let us now apply Eqs. (5-6) to the analysis of the coupled motion shown in Fig. 5-3. The initial conditions (at $t = 0$) are as follows:

$$x_A = A_0 \qquad \frac{dx_A}{dt} = 0 \qquad x_B = 0 \qquad \frac{dx_B}{dt} = 0$$

It may be noted that the conditions on the initial velocities are automatically met by Eqs. (5-6), because differentiation with respect to t gives us terms in $\sin \omega_0 t$ and $\sin \omega' t$ only, all of which go to zero at $t = 0$. From the conditions on the initial displacements themselves we have

$$\begin{aligned} x_A &= A_0 = \tfrac{1}{2}C + \tfrac{1}{2}D \\ x_B &= 0 \quad = \tfrac{1}{2}C - \tfrac{1}{2}D \end{aligned}$$

Therefore,

$$C = A_0 \qquad D = A_0$$

Hence with these particular starting conditions we have, by substitution back into equations (5-6), the following results:

$$\begin{aligned} x_A &= \tfrac{1}{2}A_0(\cos \omega_0 t + \cos \omega' t) \\ x_B &= \tfrac{1}{2}A_0(\cos \omega_0 t - \cos \omega' t) \end{aligned}$$

which can be rewritten as follows:

$$\begin{aligned} x_A &= A_0 \cos\left(\frac{\omega' - \omega_0}{2}t\right) \cos\left(\frac{\omega' + \omega_0}{2}t\right) \\ x_B &= A_0 \sin\left(\frac{\omega' - \omega_0}{2}t\right) \sin\left(\frac{\omega' + \omega_0}{2}t\right) \end{aligned} \quad (5\text{-}7)$$

Each of these is a sinusoidal oscillation of angular frequency $(\omega' + \omega_0)/2$, modulated in amplitude in the way discussed in Chapter 2. The amplitude associated with each of the pendulums is zero at the instant when the amplitude associated with the other is a maximum—although the actual *displacement* of the latter at any such instant depends on the instantaneous value of $(\omega' + \omega_0)t/2$.

OTHER EXAMPLES OF COUPLED OSCILLATORS

There are many different ways of coupling two pendulums or other oscillators together; let us consider a few.

In Fig. 5-6 we show how two pendulums may be coupled through an auxiliary mass, $m \ll M$, connected by strings to the major suspending wires. From the symmetry of the arrangement, we can guess that the normal modes will be the motions for which $x_B = \pm x_A$. If $x_A = +x_B = q_1$, the mass m rises and falls with the main masses M, but if $x_A = -x_B = q_2$, the mass m will be highest when the masses M are at their greatest separation, and will fall as the masses approach each other. Thus there are two distinct normal mode frequencies, neither of which (in general) is equal to that of one pendulum alone.

Four other mechanical coupled systems are shown in Fig. 5-7. The first diagram represents two pendulum bobs that are mounted on rigid bars, the upper ends of which are clamped to a wire. The pendulums swing in planes perpendicular to the wire. Unless the pendulums swing in phase, with equal amplitudes, the connecting wire is twisted and provides a coupling torque that is proportional to the difference of angular displacements.

In Fig. 5-7(b) we show another system in which the coupling is provided by elastic restoring forces. Two small masses are mounted at the ends of a hacksaw blade (or other strip of springy metal) which is held at its center by a yielding support. If one

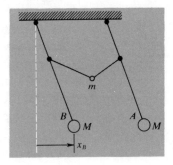

Fig. 5-6   Mass-coupled pendulums.

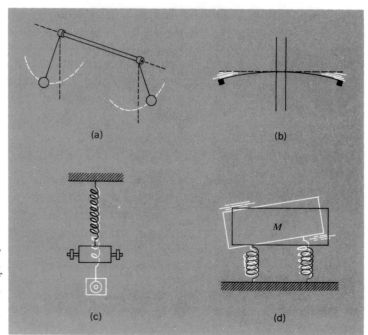

Fig. 5-7 (a) Rigid pendulums coupled by horizontal torsion rod. (b) Masses at ends of metal strip. (c) Wilberforce pendulum. (d) Rectangular block on springs.

(a)

(b)

(c)

(d)

mass is pulled aside, as shown, and then released, the motion is quickly transferred to the other mass through a typical superposition of normal modes.

Figure 5–7(c) shows a curious device known as the Wilberforce pendulum.[1] A mass with adjustable outriggers is suspended from a coil spring. If the mass is pulled down and released, the motion is at first a simple vertical oscillation, but as time goes on this oscillation dies down and is replaced by a vigorous rotational oscillation of the mass (about a vertical axis). Then the vertical linear oscillation returns as the rotational oscillation again weakens. It is important for the operation of this toy that the periods of the two types of motion be nearly equal; the adjustable outriggers are there to permit this to be arranged. The coupling between the linear and angular motions comes from the fact that, as we mentioned in Chapter 3, when a coil spring is stretched its end twists a little, or conversely that if it is twisted it tends to lengthen or shorten. By pulling the mass down and twisting it through an appropriate angle, it is possible to release the system so that it oscillates in a normal mode with constant amplitude in both components (linear and angular) of the motion.

[1]Named after L. R. Wilberforce, a British professor of physics, who published a detailed study of it in 1894.

Our last diagram [Fig. 5–7(d)] represents a rectangular block supported on two springs. One mode of this system is a vertical oscillation in which the block remains horizontal and both springs are equally stretched or compressed. But there is another mode in which the springs undergo equal and opposite displacements; the block then performs a twisting oscillation about a horizontal axis, without any change in the height of its center of gravity. A car resting on its front and rear suspensions has some resemblance to this arrangement. If the front end were lifted and then released, one might find the oscillation transferred to the rear at a later time, if damping had not already brought the system to rest.

NORMAL FREQUENCIES: GENERAL ANALYTICAL APPROACH

Suppose it were not easy to discover the normal modes from symmetry considerations, or not easy to solve the simultaneous differential equations. How then could we plough through to a solution? We make use of the characteristic we discussed in connection with Eqs. (5–6). Both x_A and x_B can oscillate with one of the normal frequencies. Let us take, therefore,

$$x_A = C \cos \omega t$$
$$x_B = C' \cos \omega t \tag{5-8}$$

and see if there are values of ω and C and C' for which these expressions are solutions of equations (5–4):

$$\frac{d^2 x_A}{dt^2} + (\omega_0{}^2 + \omega_c{}^2)x_A - \omega_c{}^2 x_B = 0$$

$$\frac{d^2 x_B}{dt^2} + (\omega_0{}^2 + \omega_c{}^2)x_B - \omega_c{}^2 x_A = 0 \tag{5-4}$$

If there are suitable values of ω, they will then be the normal frequencies. Of course, we have already found that C and C' must be equal in magnitude, but in our present approach to the problem we shall act as though we do not know that yet. Besides, the equality of C and C' is true only in the very special problem we have been considering and is not true in more general cases.

Substituting equations (5–8) into equations (5–4), we get

$$(-\omega^2 + \omega_0{}^2 + \omega_c{}^2)C \qquad\qquad - \omega_c{}^2 C' = 0$$
$$- \omega_c{}^2 C + (-\omega^2 + \omega_0{}^2 + \omega_c{}^2)C' = 0$$

For an arbitrary value of ω, these constitute two simultaneous equations for the unknown amplitudes C and C'. If they are independent equations, there is only one solution—$C = 0, C' = 0$ —which simply means that, for an arbitrary value of ω, equations (5–8) are not a solution to the problem.

But if these two equations are not independent—i.e., if the second is just a multiple of the first—then we have in effect only one equation for the two amplitudes C and C'. In this case, C can have any value. But once C is chosen, then C' is fixed.

For what value of ω are the two equations not independent and thus able to yield nonzero solutions for C and C'? From the first equation, we have

$$\frac{C}{C'} = \frac{\omega_c^2}{-\omega^2 + \omega_0^2 + \omega_c^2} \tag{5–9a}$$

and, from the second,

$$\frac{C}{C'} = \frac{-\omega^2 + \omega_0^2 + \omega_c^2}{\omega_c^2} \tag{5–9b}$$

If C and C' are not both zero, the right-hand sides of those equations must be equal. Thus

$$\frac{\omega_c^2}{-\omega^2 + \omega_0^2 + \omega_c^2} = \frac{-\omega^2 + \omega_0^2 + \omega_c^2}{\omega_c^2}$$

or

$$(-\omega^2 + \omega_0^2 + \omega_c^2)^2 = (\omega_c^2)^2$$

Hence

$$-\omega^2 + \omega_0^2 + \omega_c^2 = \pm\omega_c^2$$
$$\omega^2 = \omega_0^2 + \omega_c^2 \pm \omega_c^2$$

We have two solutions for ω; let us call them ω' and ω'':

$$\omega'^2 = \omega_0^2 + 2\omega_c^2$$
$$\omega''^2 = \omega_0^2$$

The positive square roots of these expressions are the two normal frequencies of the system; once again we have arrived at the now familiar results.

We can now get the relation between C and C' for each of the normal modes, from equations (5–9). For $\omega = \omega'$,

$$\frac{C}{C'} = -1$$

and, for $\omega = \omega''$,

$$\frac{C}{C'} = +1$$

Thus we have arrived at two specific forms of equations (5–8) which are solutions to the coupled differential equations of motion [equations (5–4)]:

$$
\begin{array}{lll}
x_A = C \cos \omega_0 t & & x_A = D \cos \omega' t \\
& \text{and} & \\
x_B = C \cos \omega_0 t & & x_B = -D \cos \omega' t
\end{array}
\qquad (5\text{–}10)
$$

Since the magnitude of the amplitude is arbitrary and determined only by the initial conditions, we have used two different symbols (i.e., C and D) to denote the amplitudes associated with the separate normal modes.

The differential equations are linear (only the first powers of x_A, x_B, d^2x_A/dt^2, and d^2x_B/dt^2 appear), and therefore the sum of the two solutions is also a solution:

$$
\text{(special case)} \qquad
\begin{array}{l}
x_A = C \cos \omega_0 t + D \cos \omega' t \\
x_B = C \cos \omega_0 t - D \cos \omega' t
\end{array}
\qquad (5\text{–}11)
$$

Once again we have obtained the solutions previously given by equations (5–6).[1] But this time our approach has been purely analytical and general, with no prior appeal to the symmetry of the system.

Let us complete this discussion by giving the general solution to the equations of free oscillation of this coupled system. It may be readily seen that the differential equations (5–4) are equally well fitted by assuming solutions with nonzero initial phases, although there *is* a systematic phase relationship between x_A and x_B in a particular mode. Specifically, instead of equations (5–10) we may in general have the following:

$$
\begin{array}{ll}
\text{Lower mode:} &
\begin{array}{l}
x_A = C \cos(\omega_0 t + \alpha) \\
x_B = C \cos(\omega_0 t + \alpha)
\end{array}
\\[2em]
\text{Higher mode:} &
\begin{array}{l}
x_A = D \cos(\omega' t + \beta) \\
x_B = -D \cos(\omega' t + \beta)
\end{array}
\end{array}
\qquad (5\text{–}12)
$$

The existence of four adjustable constants then allows us to fit these solutions to arbitrary values of the initial displacements *and* velocities of both pendulums. This removes the restriction

[1]There is a factor of 2 lacking throughout in equations (5–10) as compared with equations (5–6), but this makes no difference at all when one fixes the values of the coefficients via the initial values of x_A and x_B.

to zero initial velocity that required us to label our earlier solutions as special cases.

FORCED VIBRATION AND RESONANCE FOR TWO COUPLED OSCILLATORS

So far we have merely considered the *free* vibrations of a system of two coupled oscillators, thereby discovering the characteristic natural frequencies (just two of them) at which the system is able to vibrate as a kind of unit. But what happens if the system is driven at an arbitrary frequency by an external agency? Our intuition, backed up by actual experience, is that large amplitudes of oscillation occur when the driving frequency is close to one of the natural frequencies, whereas at frequencies far removed from these the response of the driven system is relatively small. We shall consider in detail how this emerges from the equations of motion in the simplest possible case—for two coupled identical pendulums with negligible damping, for which we have already identified the normal modes.

Our discussion will closely parallel the analysis of the forced single oscillator as in Chapter 4. Just as in that case, we shall assume that the damping effects are small enough to be ignored in the equations of motion, but that, nevertheless, perhaps after a very large number of cycles of oscillation, the transient effects have disappeared so that the motion of each pendulum occurs at constant amplitude at the frequency of the driving force.

Let us suppose, then, that a harmonic driving force $F_0 \cos \omega t$ is applied to pendulum A (e.g., by moving its point of suspension back and forth sinusoidally), the motion of pendulum B being controlled only by its own restoring force and the coupling spring. The statement of Newton's law for pendulum B is thus just the same as we had in considering the free vibrations, and the equation for A is modified only to the extent of adding the term $F_0 \cos \omega t$—although this addition represents, of course, a major change in the physical situation. Our two equations of motion thus become the following [see equations (5–3) for the free-vibration equations]:

$$m\frac{d^2 x_A}{dt^2} + m\omega_0^2 x_A + k(x_A - x_B) = F_0 \cos \omega t$$

$$m\frac{d^2 x_B}{dt^2} + m\omega_0^2 x_B - k(x_A - x_B) = 0 \qquad \left(\omega_0^2 = \frac{g}{l}\right)$$

which, dividing through by m, become

$$\frac{d^2 x_A}{dt^2} + (\omega_0{}^2 + \omega_c{}^2)x_A - \omega_c{}^2 x_B = \frac{F_0}{m}\cos \omega t$$

$$\frac{d^2 x_B}{dt^2} + (\omega_0{}^2 + \omega_c{}^2)x_B - \omega_c{}^2 x_A = 0 \qquad \left(\omega_c{}^2 = \frac{k}{m}\right)$$

Rather than dealing with x_A and x_B separately, we shall proceed at once to introduce the normal coordinates $q_1\,(= x_A + x_B)$ and $q_2\,(= x_A - x_B)$, which, as we have seen, can be used to characterize the motion of the system as a whole. Adding the differential equations above, we get

$$\frac{d^2 q_1}{dt^2} + \omega_0{}^2 q_1 = \frac{F_0}{m}\cos \omega t \tag{5-13a}$$

Subtracting them, we get

$$\frac{d^2 q_2}{dt^2} + \omega'^2 q_2 = \frac{F_0}{m}\cos \omega t \tag{5-13b}$$

where

$$\omega'^2 = \omega_0{}^2 + 2\omega_c{}^2$$

The simplification of the problem is remarkable. It is just as though we had two harmonic oscillators, of natural frequencies ω_0 and ω'. We can clearly describe the steady-state solutions by the equations

$$q_1 = C \cos \omega t \quad \text{where } C = \frac{F_0/m}{\omega_0{}^2 - \omega^2}$$

$$q_2 = D \cos \omega t \quad \text{where } D = \frac{F_0/m}{\omega'^2 - \omega^2} \tag{5-14}$$

The amplitudes C and D exhibit just the kind of resonance behavior shown for a single oscillator in Fig. 4–1. Having obtained them, we can extract the frequency dependence of the individual amplitudes A and B of the two pendulums, for we have

$$x_A = A \cos \omega t \quad \text{where} \quad A = \tfrac{1}{2}(C + D)$$

$$x_B = B \cos \omega t \quad \text{where} \quad B = \tfrac{1}{2}(C - D)$$

These give us the following results:

$$A(\omega) = \frac{F_0}{m}\frac{(\omega_0{}^2 + \omega_c{}^2) - \omega^2}{(\omega_0{}^2 - \omega^2)(\omega'^2 - \omega^2)}$$

$$B(\omega) = \frac{F_0}{m}\frac{\omega_c{}^2}{(\omega_0{}^2 - \omega^2)(\omega'^2 - \omega^2)} \tag{5-15}$$

The variation of these quantities with ω is shown in Fig. 5–8. In the region of frequencies dominated by the lower resonance, the

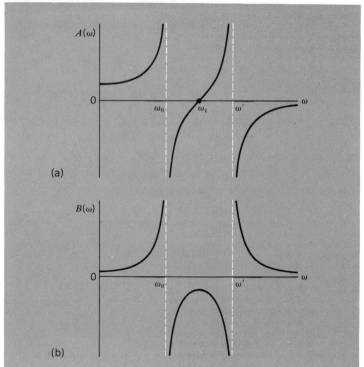

Fig. 5–8 Forced response of two coupled pendulums with negligible damping. The normal modes have the frequencies ω_0 and ω'. (a) Amplitude of first pendulum as a function of driving frequency $[\omega_1 = (\omega_0{}^2 + \omega'{}^2)^{1/2}]$. (b) Amplitude of second pendulum as a function of driving frequency.

(a)

(b)

displacements of A and B are always of the same sign—i.e., in phase with one another. In the region of frequencies dominated by the higher resonance, the displacements are of opposite sign and hence $180°$ out of phase. The introduction of nonzero damping would, as with the single driven oscillator, lead to a smooth variation of phase with frequency as one goes through the resonances.

One feature in particular of Fig. 5–8 might be commented on, because it seems (and is) physically impossible. This is the fact that at a certain frequency ω_1 between the resonances, we have $A = 0$ and B nonzero. Yet from the assumed conditions of the problem it is clear that the periodic forcing of pendulum B depends on the motion of pendulum A. In any real system some small oscillation of the bob of pendulum A would be essential. The frequency ω_1 at which the apparently anomalous situation develops is precisely the natural frequency of a single pendulum, with coupling spring attached, under the circumstance that the other pendulum is held quite fixed—$\omega_1 = (\omega_0{}^2 + \omega_c{}^2)^{1/2}$. In the complete absence of damping forces an arbitrarily small driving force of frequency ω_1, caused by arbitrarily small vibrations of pendulum A, would cause an arbitrarily large response in pendulum B. The existence of damping forces, however small, would destroy this condition, and would mean that the amplitude $A(\omega)$, although becoming very small near ω_1, would never fall quite to zero. The full description would now, however, necessitate the detailed consideration of the system as a combination of a pair of oscillators with damping, and the complexity of the analysis would be greatly increased.

The main point to be learned from this analysis is the confirmation that one can trace out the normal modes of a coupled system by means of resonance observations, and that the steady-state motions of the component parts at resonance are just like what they would be for the same system in free vibration at the same frequency.

MANY COUPLED OSCILLATORS

Any real macroscopic body, such as a piece of solid, contains many particles, not just two, so we have the strongest of motives for tackling the problem of an arbitrary number of similar oscillators coupled together. The work of the preceding sections has equipped us to do this. Our investigation of such a system can lead us to a description of the oscillations of a continuous medium, and thence by an easy transition to the analysis of wave motions.

It would be possible for us to go directly from Newton's law to continuum mechanics.[1] But the route we have chosen, via the modes of oscillation of coupled systems, is richer and in essence is more correct—for there is no such thing as a truly continuous medium. Moreover, you may be interested to know that our present route is the one that Newton and his successors themselves took. Perhaps this in itself merits an introductory digression.

Not long after Newton, two members of the remarkable Bernoulli family (John Bernoulli and his son Daniel) embarked on a detailed study of the dynamics of a line of connected masses. They showed that a system of N masses has exactly N independent modes of vibration (for motion in one dimension only). Then in 1753 Daniel Bernoulli enunciated the superposition principle for such a system—stating that the general motion of a vibrating system is describable as a superposition of its normal modes. (You will recall that earlier in this chapter we developed this

[1] As mentioned in the footnote at the beginning of this chapter, you can do this by going directly to Chapter 6.

result for the system of two oscillators.) In the words of Leon Brillouin, who has been a major contributor to the theory of crystal-lattice vibrations[1]:

> This investigation by the Bernoullis may be said to form the beginning of theoretical physics as distinct from mechanics, in the sense that it is the first attempt to formulate laws for the motion of a system of particles rather than for that of a single particle. The principle of superposition is important, as it is a special case of a Fourier series, and in time it was extended to become a statement of Fourier's theorem.

(We shall come to the notions of Fourier analysis in Chapter 6.)

After this preamble let us now turn to the detailed analysis of an N-particle system.

N COUPLED OSCILLATORS

In our treatment of the motion of a two-oscillator system, we confined our attention to oscillations which may be termed longitudinal—the motions of the pendulum bobs have been along the line connecting them. The treatment is quite similar, as we shall soon see, for transverse oscillations where the particles oscillate in a direction perpendicular to the line connecting them. And because transverse oscillations are easier to visualize and to display than longitudinal oscillations, we shall analyze the transverse oscillations of a prototype system of many particles.

Consider a flexible elastic string to which are attached N identical particles, each of mass m, equally spaced a distance l apart. Let us hold the string fixed at two points, one at a distance l to the left of the first particle and the other at a distance l to the right of the Nth particle (Fig. 5–9).

The particles are labeled from 1 to N, or from 0 to $N + 1$ if we include the two fixed ends and treat them as if they were particles with zero displacement. If the initial tension in the string is T and if we confine ourselves to small transverse displacements of the particles, then we can ignore any increase in the tension of the string as the particles oscillate. Suppose, for

Fig. 5–9 N equidistant particles along a massless string.

[1]L. Brillouin, *Wave Propagation in Periodic Structures*, Dover, New York, 1953.

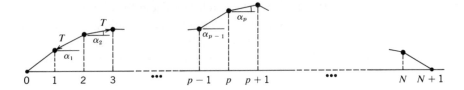

example, that particle 1 is displaced to y_1 and particle 2 to y_2 (Fig. 5–10); then the length of string between them becomes $l' = l/\cos \alpha_1$. For $\alpha_1 \ll 1$ rad, then $\cos \alpha_1 \approx 1 - \alpha_1^2/2$ and $l' \approx l(1 + \alpha_1^2/2)$. The increase in length is $l\alpha_1^2/2$, and any increased tension that is proportional to this may be ignored in comparison to any term proportional to the first power of α_1.

In the configuration as shown the resultant x component of force on particle 2 is $-T \cos \alpha_1 + T \cos \alpha_2 = \frac{1}{2}T(\alpha_1^2 - \alpha_2^2)$, a difference between two second-power terms in α. For small values of α_1 and α_2, it is exceedingly small and we shall pay it no attention in what follows.

Figure 5–10 shows a configuration of the particles at some instant of time during their transverse motion. We shall restrict ourselves to y displacements that are small compared to l. The resultant y component of force on a typical particle, say the pth particle, is

$$F_p = -T \sin \alpha_{p-1} + T \sin \alpha_p$$

The approximate values of the sines are

$$\sin \alpha_{p-1} = \frac{y_p - y_{p-1}}{l}$$

$$\sin \alpha_p = \frac{y_{p+1} - y_p}{l}$$

Therefore,

$$F_p = -\frac{T}{l}(y_p - y_{p-1}) + \frac{T}{l}(y_{p+1} - y_p)$$

and this must equal the mass m times the transverse acceleration of the pth particle. Thus

$$\frac{d^2 y_p}{dt^2} + 2\omega_0^2 y_p - \omega_0^2(y_{p+1} + y_{p-1}) = 0 \qquad (5\text{-}16)$$

where we have put

$$\frac{T}{ml} = \omega_0^2$$

We can write a similar equation for each of the N particles. Thus we have a set of N differential equations, one for each value of p from 1 to N. Remember that $y_0 = 0$ and $y_{N+1} = 0$.

You may find it helpful to consider the simple special cases of Eq. (5–16) for $N = 1$ and $N = 2$. If $N = 1$, we have

$$\frac{d^2 y_1}{dt^2} + 2\omega_0^2 y_1 = 0$$

There is transverse harmonic motion of angular frequency $\omega_0\sqrt{2} = (2T/ml)^{1/2}$, as one can conclude directly from a consideration of Fig. 5–11(a). If $N = 2$, we have

$$\frac{d^2 y_1}{dt^2} + 2\omega_0^2 y_1 - \omega_0^2 y_2 = 0$$

$$\frac{d^2 y_2}{dt^2} + 2\omega_0^2 y_2 - \omega_0^2 y_1 = 0$$

These are similar to Eqs. (5–4) for the two coupled pendulums, but we now have the simplification that ω_0 and ω_c are equal, so that $\omega_0^2 + \omega_c^2$ in equations (5–4) corresponds to $2\omega_0^2$ here, and ω_c^2 there becomes ω_0^2 here. The angular frequencies of the normal modes in this case are in a definite numerical relationship; their actual values are ω_0 and $\omega_0\sqrt{3}$. The modes for $N = 2$ are illustrated in Figs. 5–11(b) and (c). The actual configuration of the strings makes almost self-evident the relation between the natural frequencies here, but as we go to larger numbers of particles the results are far less obvious and we must resort to a more general type of analysis.

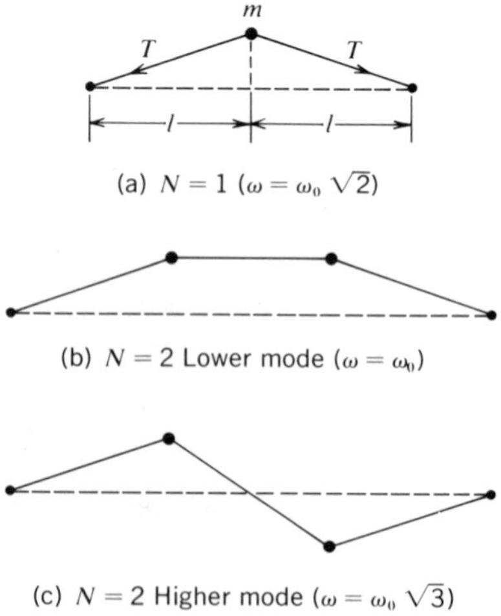

(a) $N = 1$ $(\omega = \omega_0 \sqrt{2})$

Fig. 5–11 *Normal modes of the two simplest loaded-string systems. (a) $N = 1$, one mode only. (b) $N = 2$, lower mode. (c) $N = 2$, higher mode.*

(b) $N = 2$ Lower mode $(\omega = \omega_0)$

(c) $N = 2$ Higher mode $(\omega = \omega_0 \sqrt{3})$

We apply basically the same analytical technique to our *N* differential equations as we previously used for the two equations. We seek the normal modes; i.e., we look for sinusoidal solutions such that each particle oscillates with the same frequency. We set

$$y_p = A_p \cos \omega t \qquad (p = 1, 2, \ldots, N) \qquad (5\text{-}17)$$

where A_p and ω are the amplitude and frequency of vibration of the *p*th particle. If we can find values of A_p and ω for which equations (5–17) satisfy the *N* differential equations (5–16), then we have accomplished our purpose. Note that the velocity of any particle can be obtained from equations (5–17) and is

$$\frac{dy_p}{dt} = -\omega A_p \sin \omega t \qquad (p = 1, 2, \ldots, N)$$

Thus, by choosing equations (5–17) as a trial solution, we are automatically restricting ourselves to the additional boundary condition that each particle has zero velocity at $t = 0$; i.e., each particle starts from rest.

Substituting equations (5–17) into the differential equations (5–16), we get

$$(-\omega^2 + 2\omega_0^2)A_1 - \omega_0^2(A_2 + A_0) = 0$$
$$(-\omega^2 + 2\omega_0^2)A_2 - \omega_0^2(A_3 + A_1) = 0$$
$$\vdots$$
$$(-\omega^2 + 2\omega_0^2)A_p - \omega_0^2(A_{p+1} + A_{p-1}) = 0$$
$$\vdots$$
$$(-\omega^2 + 2\omega_0^2)A_N - \omega_0^2(A_{N+1} - A_{N-1}) = 0$$

This formidable-looking set of *N* simultaneous equations can be written more compactly as follows:

$$(-\omega^2 + 2\omega_0^2)A_p - \omega_0^2(A_{p-1} + A_{p+1}) = 0$$
$$(p = 1, 2, \ldots, N) \qquad (5\text{-}18)$$

Our earlier boundary condition requiring the ends to be held fixed means that $A_0 = 0$ and $A_{N+1} = 0$.

The question we are asking ourselves is whether all *N* of these equations can be satisfied by using the same value of ω^2 in each. We saw earlier how to tackle such a problem when only two coupled oscillators were involved. The assumption that a solution existed (other than the trivial one of having all amplitudes equal to zero) led to restrictions on the ratios of the amplitudes [as expressed by equations (5–9)]. We have the same situation in this more complex problem. If we rewrite equations (5–18) as

$$\frac{A_{p-1} + A_{p+1}}{A_p} = \frac{-\omega^2 + 2\omega_0^2}{\omega_0^2} \qquad (p = 1, 2, \ldots, N) \qquad (5\text{–}19)$$

we see that, for any particular value of ω, the right side is constant, and therefore the ratio on the left must be a constant and independent of the value of p. What values can be assigned to the A_p's such that this condition will be satisfied and at the same time give $A_0 = 0$ and $A_{N+1} = 0$?

We shall not pretend to *solve* Eq. (5–19) but will simply draw attention to a remarkable result that gives the key to the problem. Suppose that the amplitude of particle p is expressible in the form

$$A_p = C \sin p\theta \qquad (5\text{–}20)$$

where θ is some angle. If a similar equation is used to define the amplitudes of the adjacent particles $p - 1$ and $p + 1$, we shall have

$$A_{p-1} + A_{p+1} = C[\sin(p - 1)\theta + \sin(p + 1)\theta]$$
$$= 2C \sin p\theta \cos \theta$$

But $C \sin p\theta$ is just A_p, so that we have

$$\frac{A_{p-1} + A_{p+1}}{A_p} = 2 \cos \theta \qquad (5\text{–}21)$$

This means that the recipe represented by Eq. (5–20) is successful. The right-hand side of Eq. (5–21) is a constant, independent of p, which is just what we need so as to have a condition equivalent to Eq. (5–19). It can be used to satisfy all N of the equations (5–18) from which we started. All that remains is to find the value of θ. This we can do by imposing the requirement that $A_p = 0$ for $p = 0$ and $p = N + 1$. The former condition is automatically satisfied; the latter will hold good if $(N + 1)\theta$ is set equal to any integral multiple of π. Thus we put

$$(N + 1)\theta = n\pi \qquad (n = 1, 2, 3, \ldots)$$
$$\theta = \frac{n\pi}{N + 1} \qquad (5\text{–}22)$$

Substituting for θ in Eq. (5–20) we thus get

$$A_p = C \sin\left(\frac{pn\pi}{N + 1}\right) \qquad (5\text{–}23)$$

The permitted frequencies of the normal modes are also determined, for from Eqs. (5–19) through (5–22) we have

$$\frac{A_{p+1} + A_{p-1}}{A_p} = \frac{-\omega^2 + 2\omega_0^2}{\omega_0^2} = 2\cos\left(\frac{n\pi}{N+1}\right)$$

Therefore,

$$\omega^2 = 2\omega_0^2\left[1 - \cos\left(\frac{n\pi}{N+1}\right)\right]$$

$$= 4\omega_0^2 \sin^2\left[\frac{n\pi}{2(N+1)}\right]$$

Taking the square root of this, we have

$$\omega = 2\omega_0 \sin\left[\frac{n\pi}{2(N+1)}\right] \qquad (5\text{-}24)$$

PROPERTIES OF THE NORMAL MODES FOR N COUPLED OSCILLATORS

Having obtained the mathematical solutions to this problem of N coupled oscillators, let us look more closely at the motions that the equations describe.

First, we observe that, according to Eq. (5–24), different values of the integer n define different normal mode frequencies. It is therefore appropriate to label a mode, and its distinctive frequency, by the value of n. Thus we shall put

$$\omega_n = 2\omega_0 \sin\left[\frac{n\pi}{2(N+1)}\right] \qquad (5\text{-}25)$$

Next, we must recognize that the motion of a given particle (or oscillator) depends both on its number along the line (p) and on the mode number (n). The amplitude of its motion can thus be written as follows:

$$A_{pn} = C_n \sin\left(\frac{pn\pi}{N+1}\right) \qquad (5\text{-}26)$$

where C_n defines the amplitude with which the particular mode n is excited. The actual displacement of the pth particle when the entire collection of particles is oscillating in the nth mode is thus given by

$$y_{pn}(t) = A_{pn} \cos \omega_n t \qquad (5\text{-}27)$$

where ω_n and A_{pn} are given by Eqs. (5–25) and (5–26), respectively. The above equation implies that each particle is at rest at the time $t = 0$, but as with the two-oscillator problem we can satisfy arbitrary initial conditions by putting

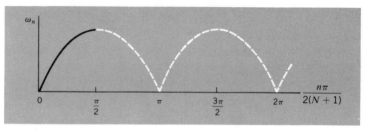

Fig. 5–12 Graph of the mode frequency as a function of mode number. It is convenient to graph ω_n against the quantity $n\pi/2(N + 1)$ rather than against n itself.

$$y_{pn}(t) = A_{pn} \cos(\omega_n t - \delta_n) \qquad (5\text{-}27a)$$

where each different mode can be assigned its own phase δ_n.

How many normal modes are there? We saw that with two coupled oscillators there were just two normal modes. If your intuition should tell you that with N oscillators there are only N independent modes, you would be right.[1] This fact is, however, somewhat hidden in Eqs. (5–25) and (5–26), because values of ω_n and A_{pn} are defined for every integral value of n. The point is, though, that beyond $n = N$ the equations do not describe any physically new situations.

We can make this clear, as far as the mode frequencies are concerned, with the help of Fig. 5–12. This is a graph of Eq. (5–25) —modified to the extent that ω is defined as being always positive. As we go from $n = 1$ to $n = N$ we find N different characteristic frequencies. At $n = N + 1$, which corresponds to $\pi/2$ on the abscissa, a maximum frequency $\omega_{\max} (= 2\omega_0)$ is reached, but it does not correspond to a possible motion because [as Eq. (5–26) shows] all the amplitudes A_{pn} are zero at this value of n. For $n = N + 2$, we have

$$\omega_{N+2} = 2\omega_0 \sin\left[\frac{(N + 2)\pi}{2(N + 1)}\right]$$

$$= 2\omega_0 \sin\left[\pi - \frac{N\pi}{2(N + 1)}\right]$$

$$= 2\omega_0 \sin\left[\frac{N\pi}{2(N + 1)}\right]$$

Therefore,

$$\omega_{N+2} = \omega_N$$

Similarly, $\omega_{N+3} = \omega_{N-1}$, and so on. And a similar duplication occurs in every subsequent range of $N + 1$ values of n.

[1]This is for a one-dimensional system. Two dimensions gives $2N$, three dimensions gives $3N$.

142   Coupled oscillators and normal modes

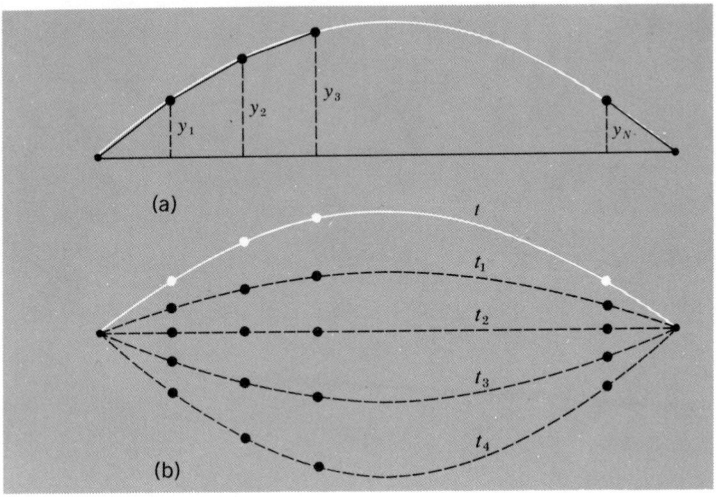

Fig. 5–13 (a) Plot
of sin [pπ/(N + 1)]
as a function of p.
The particles are at
the positions defined
by integral values of p
and are joined by
straight segments of
string. (b) Positions
of particles at various
times for lowest mode.

It is only a short step to see that the relative amplitudes of the particles in a normal mode repeat themselves also. Thus, for example, we have, from Eq. (5–26),

$$A_{p,N+2} = C_{N+2} \sin\left[\frac{p(N + 2)\pi}{N + 1}\right]$$

$$= C_{N+2} \sin\left[2p\pi - \frac{pN\pi}{N + 1}\right]$$

$$= -C_{N+2} \sin\left(\frac{pN\pi}{N + 1}\right)$$

$$\sim A_{p,N}$$

and it is easy to show that a similar matching occurs for any other $n > N + 1$.

Let us see what the various normal modes look like. The first mode is given by $n = 1$. The particle displacements are

$$y_{p1} = C_1 \sin\left(\frac{p\pi}{N + 1}\right) \cos \omega_1 t \qquad (p = 1, 2, \ldots, N)$$

At a given instant of time, the $C_1 \cos \omega_1 t$ factor is the same for all particles. Only the $\sin[p\pi/(N + 1)]$ factor distinguishes the displacements of the different particles. The white curve in Fig. 5–13(a) is a plot of $\sin[p\pi/(N + 1)]$ versus p, as p varies continuously from 0 to $N + 1$. Actual particles, however, are located at the discrete values $p = 1, 2, \ldots, N$. The sine curve is therefore only a guide for locating the particles, and the string consists of straight-line segments connecting the particles.

As t increases, each particle oscillates in the y direction with

143   Properties of modes for N coupled oscillators

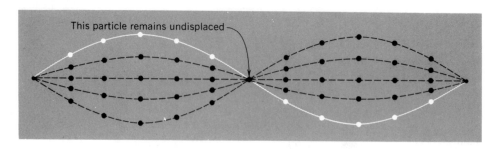

This particle remains undisplaced

Fig. 5–14   Positions of particles at various times for second mode (n = 2).

frequency ω_1. A whole set of sine curves for different values of t, and the corresponding locations of the particles, are shown in Fig. 5–13(b). For the second mode, $n = 2$ and

$$y_{p2} = C_2 \sin\left(\frac{p2\pi}{N+1}\right) \cos \omega_2 t \qquad (p = 1, 2, \ldots, N)$$

The particle displacements at different instants of time are shown in Fig. 5–14. If the number of particles should happen to be odd, there would be one particle at the center of the line and in this mode it would remain at rest, as indicated in Fig. 5–14. Remember that ω_2 differs from ω_1, and therefore this pattern oscillates with a different frequency than the previous one—almost twice as great, in fact.

In Fig. 5–15 we show a set of diagrams of the normal modes for a set of four particles on a stretched string. This displays very beautifully how the pattern of displacements retraces its steps after reaching $n = 5$, even though the sine curves that determine the A_{pn} are all different. These sketches for a small value of N also allow one to appreciate how remarkable it is that the displacements of every particle in every mode for such a system should fall upon a sine curve, when the string connecting them may follow an entirely different path.

LONGITUDINAL OSCILLATIONS

As we explained at the outset, we chose to consider transverse vibrations, rather than longitudinal ones, as a basis for analyzing the behavior of a system comprising a large number of coupled oscillators. The eye and the brain can take in, at a glance, what is happening to each and every particle when a string of masses is set into transverse oscillations. But now let us see how the

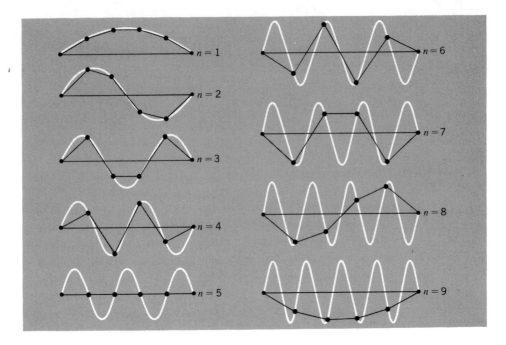

Fig. 5–15   Modes of weighted vibrating string, N = 4.
Note that n = 6, 7, 8, 9 repeat patterns of n = 4, 3,
2, 1 with opposite sign. (Adapted from J. C. Slater
and N. H. Frank, Mechanics, McGraw-Hill, New York,
1947.)

same kind of analysis applies to a system of particles connected
by springs along a straight line, and limited to motions along that
line. This may seem like a very artificial system, but a line of
atoms in a crystal is surprisingly well represented by such a model
—and so, to a lesser extent, is a column of gas.

We shall again assume that the particles are of mass m and
when at rest are spaced by distances l [Fig. 5–16(a)]. But now
the restoring forces are provided by the stretching or compression

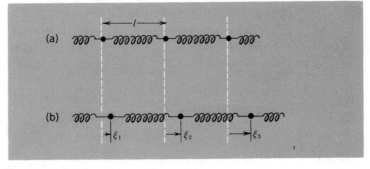

Fig. 5–16   (a) Spring-
coupled masses in
equilibrium.
(b) Spring-coupled
masses after small
longitudinal displace-
ment.

of the springs; the spring constant for each spring can be written as $m\omega_0{}^2$. Let the displacements of the masses from their equilibrium positions be denoted by $\xi_1, \xi_2, \ldots, \xi_n{}^1$ [see Fig. 5–16(b)].

Then the equation of motion of the pth particle is as follows:

$$m \frac{d^2\xi_p}{dt^2} = m\omega_0{}^2(\xi_{p+1} - \xi_p) - m\omega_0{}^2(\xi_p - \xi_{p-1})$$

i.e.,

$$\frac{d^2\xi_p}{dt^2} + 2\omega_0{}^2\xi_p - \omega_0{}^2(\xi_{p+1} + \xi_{p-1}) = 0 \tag{5-28}$$

This has precisely the same form as Eq. (5–16), so we know that mathematically all the features we have discovered for the transverse vibrations of the loaded string have their counterparts in this new system. That is to say, the motion of the pth particle in the nth normal mode is given by

$$\xi_{pn}(t) = C_n \sin\left(\frac{pn\pi}{N+1}\right) \cos \omega_n t$$

where

$$\omega_n = 2\omega_0 \sin\left[\frac{n\pi}{2(N+1)}\right] \tag{5-29}$$

A very nice quantitative study of such systems has become possible through the use of air suspensions, in which a flow of air (at pressures just a little above atmospheric) from holes in a bearing surface can be made to provide an almost completely frictionless support for objects gliding over the surface. Figure 5–17 shows the results of measurements made with such an apparatus.[2] The masses were each about 0.15 kg, and the spring constants were such that the frequency ω_0 was 5.68 sec^{-1}.

The figure shows the observed frequencies ν_n $(= \omega_n/2\pi)$ of the various normal modes, plotted as a function of the variable $n/(N+1)$. The graph contains measurements made with a system of 6 masses (and 7 springs) and with a longer but otherwise similar system of 12 masses (and 13 springs). Since ω_0 was the same for both, the results for the two systems should fall upon the single curve:

$$\nu_n = \frac{\omega_n}{2\pi} = \frac{\omega_0}{\pi} \sin\left(\frac{n}{N+1}\frac{\pi}{2}\right)$$

[1] We use the Greek letter ξ so as to reserve the ordinary x for total distance from one end.

[2] R. B. Runk, J. L. Stull, and O. L. Anderson, *Am. J. Phys.*, **31**, 915 (1963).

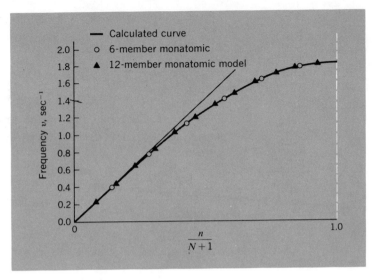

Fig. 5–17 Experimental values of mode frequency ν_n plotted against mode number for a line of identical spring-coupled masses. [Note that abscissa is $n/(N+1)$, rather than n; this allows data for two different values of N ($N = 6$ and $N = 12$) to be fitted to same theoretical curve.] [From R. B. Runk, J. L. Stull, and O. L. Anderson, Am. J. Phys., 31, 915 (1963).]

It may be seen that the experimental values conform extremely well to the theoretical ones.

N VERY LARGE

Suppose now that we allow the number of masses in a coupled system to become very large. To make the discussion explicit, we shall take the case of the transverse vibrations of particles on a stretched string. A real string, just by itself, is in fact already a collection of a large number of closely spaced atoms. Once again we can be sure that our conclusions will apply equally to the line of masses connected by springs in longitudinal vibration.

We shall let N increase but, at the same time, let the spacing l between neighboring particles decrease so that the length of string, $L = (N + 1)l$, remains constant. We shall also decrease the mass of each particle so that the total mass, $M = Nm$, also remains constant.

What happens to the normal frequencies? We have found that

$$\omega_n = 2\omega_0 \sin\left[\frac{n\pi}{2(N+1)}\right]$$

where $\omega_0 = (T/ml)^{1/2}$. First, consider the normal modes for which the mode number n is small. Then as N becomes very large, we can put

$$\sin\left[\frac{n\pi}{2(N+1)}\right] \approx \frac{n\pi}{2(N+1)}$$

Therefore,

$$\omega_n \approx 2\left(\frac{T}{ml}\right)^{1/2}\frac{n\pi}{2(N+1)} = \left(\frac{T}{m/l}\right)^{1/2}\frac{n\pi}{(N+1)l}$$

But $(N+1)l = L$, the total length of the string, and m/l is the mass per unit length (linear density) which we shall denote by μ. Thus, approximately,

$$\omega_n = n\frac{\pi}{L}\left(\frac{T}{\mu}\right)^{1/2} \qquad (n = 1, 2, \ldots) \qquad (5\text{--}30)$$

In particular,

$$\omega_1 = \frac{\pi}{L}\left(\frac{T}{\mu}\right)^{1/2}$$

and then $\omega_n = n\omega_1$. The normal frequencies are integral multiples of the lowest frequency ω_1. Remember, however, that this *is* only an approximation, even though for $n \ll N$ it is an exceedingly good one.

What about the particle displacements? Previously, we found that, in the nth mode, the displacement of the pth particle is

$$y_{pn} = C_n \sin\left(\frac{pn\pi}{N+1}\right)\cos\omega_n t$$

Instead of denoting the particle by its p value, we can specify its distance, x, from the fixed end of the string. Now

$$x = pl$$

Hence

$$\frac{pn\pi}{N+1} = \frac{pln\pi}{(N+1)l} = \frac{n\pi x}{L}$$

In place of y_{pn}, we can write $y_n(x, t)$, by which we mean the y displacement at the time t of the particle located at x, when the string is vibrating in the nth mode. Thus

$$y_n(x, t) = C_n \sin\left(\frac{n\pi x}{L}\right)\cos\omega_n t \qquad (n = 1, 2, \ldots) \qquad (5\text{--}31)$$

As N becomes very large, the x values, which locate the particles, get closer and closer together and x can be taken as a continuous variable going from 0 to L. The white sine curves of Figs. 5–13, 5–14, and 5–15 are now the actual configurations of the string in its different modes. It does not take much imagination to

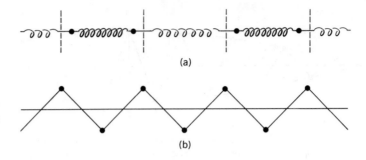

Fig. 5–18 (a) Longitudinal vibrations in the highest mode of a line of spring-coupled masses. (b) Transverse vibrations in the highest mode of a line of masses on a stretched string.

connect such motions with the possibility of wave disturbances traveling along the string, but we shall not proceed to that subject just yet.

Let us now consider the *highest* possible mode, $n = N$. If N is very large, we have

$$\omega_{max} = 2\omega_0 \sin\left[\frac{N\pi}{2(N+1)}\right] \approx 2\omega_0 \sin\left(\frac{\pi}{2}\right) = 2\omega_0 \qquad (5\text{-}32)$$

In this mode (as we shall show in a moment) each particle has, at every instant, a displacement that is opposite in sign to the displacements of its nearest neighbors, and—except for those particles near to one or the other of the fixed ends—these displacements are almost equal in magnitude. Thus for longitudinal oscillations the situation is somewhat as indicated in Fig. 5–18(a), and for the more readily visualizable case of transverse oscillations it is like Fig. 5–18(b).

This relationship of the adjacent displacements can be inferred with the help of Eq. (5–26):

$$A_{pn} = C_n \sin\left(\frac{pn\pi}{N+1}\right)$$

Putting $n = N$, we have

$$A_{p,N} = C_N \sin\left(\frac{pN\pi}{N+1}\right)$$

which we can write as

$$A_{p,N} = C_N \sin(p\pi - \alpha_p)$$

where

$$\alpha_p = \frac{p\pi}{N+1}$$

First, note that in going from p to $p+1$, the sign of the amplitude is reversed, because the angle $p\pi$ changes from an odd to an even multiple of π (or vice versa) and the angle α_p is less than π for

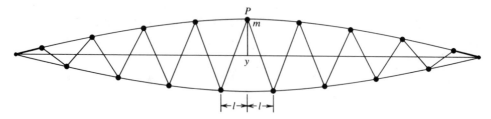

Fig. 5–19 *Amplitudes of a complete line of particles in the highest mode for a string fixed at both ends.*

every p (since $p \leq N$). This puts successive values of $(p\pi - \alpha_p)$ into opposite quadrants. Thus we can put

$$\text{(highest mode, } n = N) \quad \frac{A_p}{A_{p+1}} = -\frac{\sin\left[\dfrac{p\pi}{N+1}\right]}{\sin\left[\dfrac{(p+1)\pi}{N+1}\right]} \quad (5\text{-}33)$$

Notice next that, apart from the alternation of sign, Eq. (5–33) describes a distribution of amplitudes that fit on a half-sine curve drawn between the two fixed ends, as shown in Fig. 5–19 for the case of transverse vibrations of a line of masses.[1] Thus over most of the central region of the line the displacements are almost equal and opposite. Consider, for example, a line of 1000 masses. Then for $100 \leq p \leq 900$ the successive amplitudes differ by less than 1%. It is only toward the ends of the line that the appearance differs markedly from Fig. 5–18(b). It is then easy to see why the frequency should be nearly equal to $2\omega_0$. Consider the particle P in Fig. 5–19. If its displacement at some instant is y, the displacements of its neighbors are both approximately $-y$. Thus if the tension in the connecting strings is T, the transverse component of force due to each is approximately $(2y/l)T$, and the equation of motion of P is given by

$$m\frac{d^2y}{dt^2} \approx -2T\frac{2y}{l}$$

or

$$\frac{d^2y}{dt^2} \approx -\frac{4T}{ml}y = -4\omega_0{}^2 y$$

(Remember that the magnitudes of the transverse displacements are grossly exaggerated in the diagrams; we really are supposing $y \ll l$, as usual.) The above equation thus defines SHM of angular

[1]Note that this result holds for the highest mode even for small N—see, for example, the fourth diagram in Fig. 5–15.

frequency $2\omega_0$ approximately—and a little further consideration will convince you that the exact frequency is a shade *less* than $2\omega_0$, just as Eq. (5–32) requires.

In all of our discussion of normal modes up until now we have, with good reason, laid great emphasis on the boundary conditions that are applied—whether, for example, the ends of a line of masses are fixed or free. It may, however, have become apparent to you during this last discussion that the properties of the very high modes of a line of very many particles depend relatively little on the precise boundary conditions, even though the low modes are critically dependent on them. Thus the above calculation of the highest mode frequency of the system requires only the realization that the displacements of successive particles are approximately equal and opposite. We should have arrived at the same approximate value of the highest mode frequency if we had assumed that one end of the line was fixed and the other end free. It should be realized, however, that this *is* only approximately true, and that the effect of the precise boundary conditions must always in principle be considered.

NORMAL MODES OF A CRYSTAL LATTICE

We shall not do more than touch on this subject, which, in fact, requires whole books to do it justice. However, the analysis of the previous section carries over in a very successful way to the description of the vibrational modes of solids. This is not too surprising, because, as we have remarked, the interaction between adjacent atoms is, as far as small displacements are concerned, remarkably like that of a spring. And the structure of a solid is a lattice of greater or lesser regularity, justifying the frequently used comparison of a crystal lattice to a three-dimensional bedspring with respect to its vibrational behavior.

If we try to apply Eqs. (5–29) and (5–30) to a solid, we can think of a line of atoms along one of the principal directions in the lattice, so that μ is the total mass of all the atoms per unit length, or the mass of one atom divided by the interatomic separation, l. But what is the tension T? In Chapter 3 we introduced a strong hint for calculating the spring constant due to internal elastic forces. Dimensionally, the ratio T/μ is the same as the ratio Y/ρ of the Young's modulus to the density. The use of this is suggested even more strongly when we think of stretched springs as shown in Fig. 5–16. Thus we shall consider the possibility of describing crystal vibration frequencies $\nu\ (= \omega/2\pi)$ through the following relation:

$$\nu_n = 2\nu_0 \sin\left[\frac{n\pi}{2(N+1)}\right] \quad \text{where } \nu_0 = \frac{1}{2l}\left(\frac{Y}{\rho}\right)^{1/2} \tag{5–34}$$

For solids, as we have seen (see Table 3–1), the values of Y are of the order of 10^{11} N/m^2, so that, because the densities ρ are of the order of 10^4 kg/m^3, the ratio Y/ρ is of the order of 10^7 m^2/sec^2. The interatomic distance l is of the order of 10^{-10} m. Thus we should have

$$\nu_0 \approx 10^{13} \text{ sec}^{-1}$$

This is the highest frequency that the lattice could support. The low modes are well described by the analogues of Eq. (5–30):

$$\nu_n = \frac{1}{2L}\left(\frac{Y}{\rho}\right)^{1/2}$$

where L is the thickness of the crystal. Thus the *lowest* frequency of vibration of a crystal 1 cm across would be of the order of 10^5 Hz.

To return to the highest possible mode, this is the one in which adjacent atoms are displaced oppositely to one another (see Fig. 5–18). Such motion can be very effectively stimulated by light falling upon an ionic crystal such as sodium chloride, in which the Na$^+$ and Cl$^-$ ions are always being pushed in opposite directions by the electric field of the light wave. From our very rough calculation, we see that a resonance condition between the light and the lattice might be expected to occur at a frequency of the order of 10^{13} Hz, corresponding to a wavelength of the order of 3×10^{-5} m, or 30μ. This is infrared. Figure 5–20

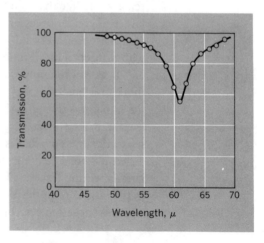

Fig. 5–20 Transmission of infrared radiation through a thin (0.17 μ) sodium chloride film. [After R. B. Barnes, Z. Physik, 75, 723 (1932).]

152 Coupled oscillators and normal modes

shows a beautiful example of just such a resonance, resulting in increased absorption of light by the crystal at wavelengths in the neighborhood of 60μ. It was observed using an extremely thin slice of NaCl—only about 10^{-7} m thick.

PROBLEMS

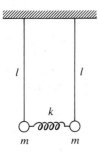

5-1 The best way to get a feeling for the behavior of a coupled oscillator system is to make your own, and experiment with it under various conditions. Try making a pair of identical pendulums, connected by a drinking straw that can be set at various distances down the threads (see sketch). Study the motions for oscillations both in the plane of the pendulums (when they move toward or away from one another) and also perpendicular to this plane. Try measuring the normal mode periods and also the period of transfer of motion from one to the other and back. Do your results conform to what the text describes?

5-2 Two identical pendulums are connected by a light coupling spring. Each pendulum has a length of 0.4 m, and they are at a place where $g = 9.8$ m/sec^2. With the coupling spring connected, one pendulum is clamped and the period of the other is found to be 1.25 sec exactly.

   (a) With neither pendulum clamped, what are the periods of the two normal modes?

   (b) What is the time interval between successive maximum possible amplitudes of one pendulum after one pendulum is drawn aside and released?

5-3 A mass m hangs on a spring of spring constant k. In the position of static equilibrium the length of the spring is l. If the mass is drawn sideways and then released, the ensuing motion will be a combination of (a) pendulum swings and (b) extension and compression of the spring. Without using a lot of mathematics, consider the behavior of this arrangement as a coupled system.

5-4 Two harmonic oscillators A and B, of mass m and spring constants k_A and k_B, respectively, are coupled together by a spring of spring constant k_C. Find the normal frequencies ω' and ω'' and describe the normal modes of oscillation if $k_C{}^2 = k_A k_B$.

5-5 Two identical undamped oscillators, A and B, each of mass m and natural (angular) frequency ω_0, are coupled in such a way that the coupling force exerted on A is $\alpha m(d^2 x_B/dt^2)$, and the coupling force exerted on B is $\alpha m(d^2 x_A/dt^2)$, where α is a coupling constant of magnitude less than 1. Describe the normal modes of the coupled system and find their frequencies.

5-6 Two equal masses on an effectively frictionless horizontal air track are held between rigid supports by three identical springs, as shown. The displacements from equilibrium along the line of the springs are described by coordinates x_A and x_B, as shown. If either of the masses is clamped, the period T $(= 2\pi/\omega)$ for one complete vibration of the other is 3 sec.

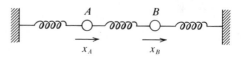

(a) If both masses are free, what are the *periods* of the two normal modes of the system? Sketch graphs of x_A and x_B versus t in each mode. At $t = 0$, mass A is at its normal resting position and mass B is pulled aside a distance of 5 cm. The masses are released from rest at this instant.

(b) Write an equation for the subsequent displacement of each mass as a function of time.

(c) What length of time (in seconds) characterizes the periodic transfer of the motion from B to A and back again? After one cycle, is the situation at $t = 0$ exactly reproduced? Explain.

5-7 Two objects, A and B, each of mass m, are connected by springs as shown. The coupling spring has a spring constant k_c, and the other two springs have spring constant k_0. If B is clamped, A vibrates at a frequency ν_A of 1.81 sec^{-1}. The frequency ν_1 of the lower normal mode is 1.14 sec^{-1}.

(a) Satisfy yourself that the equations of motion of A and B are

$$m\frac{d^2x_A}{dt^2} = -k_0x_A - k_c(x_A - x_B)$$

$$m\frac{d^2x_B}{dt^2} = -k_0x_B - k_c(x_B - x_A)$$

(b) Putting $\omega_0 = \sqrt{k_0/m}$, show that the angular frequencies ω_1 and ω_2 of the normal modes are given by

$$\omega_1 = \omega_0, \qquad \omega_2 = [\omega_0{}^2 + (2k_c/m)]^{1/2},$$

and that the angular frequency of A when B is clamped ($x_B = 0$ always) is given by

$$\omega_A = [\omega_0{}^2 + (k_c/m)]^{1/2}$$

(c) Using the numerical data above, calculate the expected frequency (ν_2) of the higher normal mode. (The observed value was 2.27 sec^{-1}.)

(d) From these same data calculate the ratio k_c/k_0 of the two spring constants.

5-8 (a) A force F is applied at point A of a pendulum as shown. At what angle θ ($\ll 1$ rad) is the new equilibrium position? What force F', applied at m, would produce the same result?

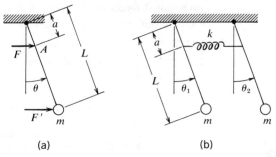

(a)   (b)

Two identical pendulums consisting of equal masses mounted on rigid, weightless rods, are arranged as shown. A light spring (unstretched when both rods are vertical, and placed as shown) provides the coupling.

(b) Write down the differential equations of motion for *small-amplitude* oscillations in terms of θ_1 and θ_2. (Neglect damping.)

(c) Describe the motion of the pendulums in each of the normal modes.

(d) Calculate the frequencies of the normal modes of the system.

[*Hint:* The symmetry of the system can be exploited to good advantage, particularly in parts (c) and (d), as long as the answers obtained this way are checked in the equations.]

5-9 The CO_2 molecule can be likened to a system made up of a central mass m_2 connected by equal springs of spring constant k to two masses m_1 and m_3 (with $m_3 = m_1$).

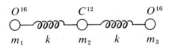

(a) Set up and solve the equations for the two normal modes in which the masses oscillate along the line joining their centers. [The equation of motion for m_3 is $m_3(d^2x_3/dt^2) = -k(x_3 - x_2)$ and similar equations can be written for m_1 and m_2.]

(b) Putting $m_1 = m_3 = 16$ units, $m_2 = 12$ units, what would be the ratio of the frequencies of the two modes, assuming this classical description were applicable?

5-10 Two equal masses are connected as shown with two identical massless springs of spring constant k. Considering only motion in the vertical direction, show that the angular frequencies of the two normal modes are given by $\omega^2 = (3 \pm \sqrt{5})k/2m$ and hence that the ratio of the normal mode frequencies is $(\sqrt{5} + 1)/(\sqrt{5} - 1)$. Find the ratio of amplitudes of the two masses in each separate mode. (*Note:*

You need not consider the gravitational forces acting on the masses, because they are independent of the displacements and hence do not contribute to the restoring forces that cause the oscillations. The gravitational forces merely cause a shift in the equilibrium positions of the masses, and you do not have to find what those shifts are.)

5–11 The sketch shows a mass M_1 on a frictionless plane connected to support O by a spring of stiffness k. Mass M_2 is supported by a string of length l from M_1.

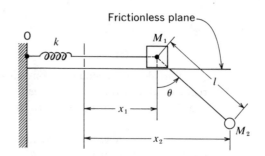

(a) Using the approximation of small oscillations,

$$\sin \theta \approx \tan \theta \approx \frac{x_2 - x_1}{l}$$

and starting from $F = ma$, derive the equations of motion of M_1 and M_2:

$$M_1 \ddot{x}_1 = -kx_1 + M_2 \frac{g}{l} (x_2 - x_1)$$

$$M_2 \ddot{x}_2 = - \frac{M_2 g}{l} (x_2 - x_1)$$

(b) For $M_1 = M_2 = M$, use the equations to obtain the normal frequencies of the system.

(c) What are the normal-mode motions for $M_1 = M_2 = M$ and $g/l \gg k/M$?

5–12 Two equal masses m are connected to three identical springs (spring constant k) on a frictionless horizontal surface (see figure). One end of the system is fixed; the other is driven back and forth with a displacement $X = X_0 \cos \omega t$. Find and sketch graphs of the resulting displacements of the two masses.

5–13 A string of length $3l$ and negligible mass is attached to two fixed supports at its ends. The tension in the string is T.

(a) A particle of mass m is attached at a distance l from one end of the string, as shown. Set up the equation for small transverse oscillations of m, and find the period.

(b) An additional particle of mass m is connected to the string as shown, dividing it into three equal segments each with tension T. Sketch the appearance of the string and masses in the two separate normal modes of transverse oscillations.

(c) Calculate ω for that normal mode which has the higher frequency.

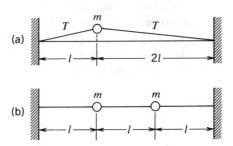

5–14 To get a feeling for the use of the equation,

$$A_{pn} = C_n \sin\left(\frac{pn\pi}{N+1}\right)$$

[Eq. (5–26) in the text], which describes the amplitudes of connected particles in the various normal modes, take the case $N = 3$ and tabulate, in a 3×3 array, the relative numerical values of the amplitudes of the particles ($p = 1, 2, 3$) in each of the normal modes ($n = 1, 2, 3$).

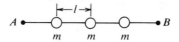

5–15 An elastic string of negligible mass, stretched so as to have a tension T, is attached to fixed points A and B, a distance $4l$ apart, and carries three equally spaced particles of mass m, as shown.

(a) Suppose that the particles have small transverse displacements y_1, y_2, and y_3, respectively, at some instant. Write down the differential equation of motion for each mass.

(b) The appearance of the normal modes can be found by drawing the sine curves that pass through A and B. Sketch such curves so as to find the relative values and signs of A_1, A_2, and A_3 in each of the possible modes of the system.

(c) Putting $y_1 = A_1 \sin \omega t$, $y_2 = A_2 \sin \omega t$, $y_3 = A_3 \sin \omega t$ in the equations (a), use the ratios $A_1:A_2:A_3$ from part (b) to find the angular frequencies of the separate modes.

5-16 Consider a system of N coupled oscillators driven at a frequency $\omega < 2\omega_0$ (i.e., $y_0 = 0$, $y_{N+1} = h \cos \omega t$). Find the resulting amplitudes of the N oscillators. [*Hint:* The differential equations of motion are the same as in the undriven case (only the boundary conditions are different). Hence try $A_p = C \sin \alpha p$, and determine the necessary values of α and C. (*Note:* If $\omega > 2\omega_0$, α is complex and the wave damps exponentially in space.)]

5-17 It is shown in the text that the highest normal-mode frequency of a line of masses can be found by considering a particle near the middle of the line, bordered by particles that have almost equal and opposite displacements to its own. Show that the same frequency can be calculated by considering the *first* particle in the line, acted on by the tension in the segments of string joining it to the fixed end and to particle 2 (see Fig. 5–19 and the related discussion).

Here we are concerned with one of the most ancient branches of mathematics, the theory of the vibrating string, which has its roots in the ideas of the Greek mathematician Pythagoras.

NORBERT WIENER, *I Am a Mathematician* (1956)

6

Normal modes of continuous systems. Fourier analysis

OUR DISCUSSIONS in this chapter will not be limited to vibrating strings. If they were, one might well question their importance. After all, who, apart from a segment of the musicians' community, depends on stretched strings for making a living? The fact is, though, that through a full analysis of this almost absurdly elementary physical system—through an understanding of its dynamics, its natural vibrations, its response at different frequencies—we are introduced to results and concepts that have their counterparts throughout the realm of physics, including electromagnetic theory, quantum mechanics, and all the rest. We are not primarily concerned with studying the string for its own sake, but it provides an almost ideal starting point. In particular, as far as mechanics proper is concerned, we can proceed from the analysis of the string to the vibrational behavior of almost any system that can be regarded as having a continuous structure. Ultimately, as we know, on a sufficiently microscopic scale this analysis must fail; we shall be driven back to the picture of any piece of material as being made up of great numbers of discrete particles, strongly interacting with one another. That was the subject of Chapter 5. But any piece of ordinary matter, large enough to be seen or touched, is so nearly homogeneous and continuous that it is profitable, and for most purposes justifiable,

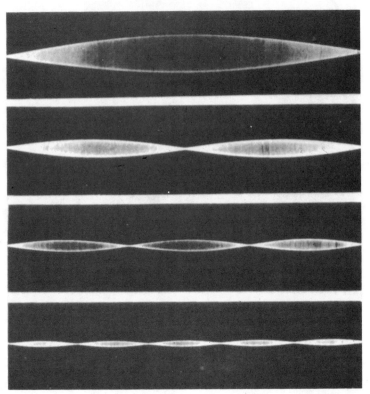

Fig. 6-1 Vibration of a string in various simple modes (n = 1, 2, 3, 5). (From D. C. Miller, The Science of Musical Sounds, *Macmillan, New York, 1922.)*

to make a fresh analysis of its behavior from this macroscopic point of view. That, then, will be the basis of everything we do in the present chapter.

THE FREE VIBRATIONS OF STRETCHED STRINGS

As implied by the quotation at the beginning of this chapter, the study of vibrating strings has a long history. The reason is, of course, the musical use of stretched strings since time immemorial. Pythagoras is said to have observed how the division of a stretched string into two segments gave pleasing sounds if the lengths of those segments bore a numerically simple ratio. Our interest here, however, is not in the musical effects, but in the basic mechanical fact that a string, with both ends fixed, has a number of well-defined states of natural vibration, as shown in Fig. 6–1. These are called *stationary vibrations*, in the sense that each point on the string vibrates transversely in SHM with constant amplitude, the frequency of this vibration being the same for all parts of the string. Such stationary vibrations represent the so-called *normal*

162 Normal modes of continuous systems

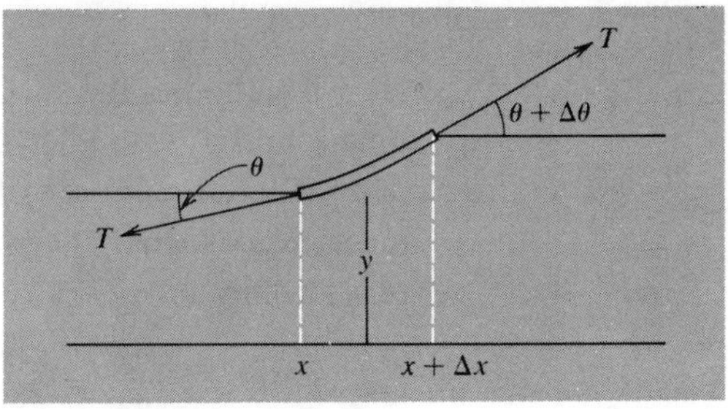

Fig. 6-2 Force diagram for short segment of massive string in transverse vibration.

modes of the string. In all except the lowest mode, there exist points at which the displacement remains zero at all times. These are *nodes*; the positions of maximum amplitude are called *antinodes*. One can thus think of these basic states of vibration as being stationary in the additional sense that the nodes remain at fixed points along the string. This is made especially clear in Fig. 6-1, because it is a time exposure.

Let us now consider the dynamics of such vibrations. We shall suppose that the string is of length L, with its ends held fixed at the points $x = 0$, $x = L$. We shall suppose further that the string has a uniform linear density (mass per unit length) equal to μ and that it is stretched with a tension T.[1] At some instant let the configuration of some portion of the string be as shown in Fig. 6-2. In Chapter 5 we considered the equivalent problem for point particles connected by a massless string, and showed then that to a good approximation the tension remains unchanged when the system is deformed from its equilibrium configuration. Thus, for a short segment of the string, of length Δx, the net force acting on it is given by

$$F_y = T\sin(\theta + \Delta\theta) - T\sin\theta$$
$$F_x = T\cos(\theta + \Delta\theta) - T\cos\theta$$

where θ, $\theta + \Delta\theta$ are the directions of tangents to the string at the ends of the segment, i.e., at x and $x + \Delta x$.

We are assuming that the transverse displacement y is small, so that θ and $\theta + \Delta\theta$ are small angles. In this case we have

$$F_y \approx T\Delta\theta$$
$$F_x \approx 0$$

The equation governing the transverse motion of the segment is thus (very nearly)

[1] In this chapter the symbol T will *not* be used to denote the period of a vibration.

163   The free vibrations of stretched strings

$$T \Delta\theta = (\mu \Delta x)a_y \tag{6-1}$$

Now θ embodies the variation of y with x at a given value of the time t, and a_y embodies the variation of y with t at a given value of x. Therefore, in rewriting Eq. (6-1) in terms of x, y, and t we must use partial derivatives, and we have the following relationship:

$$\tan \theta = \frac{\partial y}{\partial x}$$

$$\sec^2 \theta \, \Delta\theta = \frac{\partial^2 y}{\partial x^2} \Delta x$$

But $\sec \theta \approx 1$, and so

$$\Delta\theta \approx \frac{\partial^2 y}{\partial x^2} \Delta x$$

Also

$$a_y = \frac{\partial^2 y}{\partial t^2}$$

Thus Eq. (6-1) becomes

$$T\frac{\partial^2 y}{\partial x^2} \Delta x = \mu \, \Delta x \frac{\partial^2 y}{\partial t^2}$$

Therefore,

$$\frac{\partial^2 y}{\partial x^2} = \frac{\mu}{T} \frac{\partial^2 y}{\partial t^2} \tag{6-2}$$

It is clear from this equation that T/μ has the dimension of the square of a speed, and this will prove to be none other than the speed with which *progressive* waves travel on a long string having these values of μ and T. This aspect of things, however, we shall not take up until Chapter 7. For the moment, we shall simply define the speed v through the equation

$$v = \left(\frac{T}{\mu}\right)^{1/2} \tag{6-3}$$

and will then rewrite Eq. (6-2) in the following more compact form:

$$\frac{\partial^2 y}{\partial x^2} = \frac{1}{v^2} \frac{\partial^2 y}{\partial t^2} \tag{6-4}$$

We shall now look for solutions of this equation corresponding to the kind of situation physically represented by a stationary vibration. This means that every point on the string is moving with a time dependence of the form $\cos \omega t$, but that the amplitude of this motion is a function of the distance x of that point from the end of the string. (Our assumed time dependence would require every point on the string to be instantaneously stationary at $t = 0$. If this is not so an initial phase angle must be introduced.)

Thus we assume

$$y(x, t) = f(x) \cos \omega t \tag{6-5}$$

This then gives us

$$\frac{\partial^2 y}{\partial t^2} = -\omega^2 f(x) \cos \omega t$$

$$\frac{\partial^2 y}{\partial x^2} = \frac{d^2 f}{dx^2} \cos \omega t$$

(Notice that, since f is by definition a function of x only, we can write $d^2 f / dx^2$, instead of a partial derivative.) Substituting these derivatives in Eq. (6-4) then gives us

$$\frac{d^2 f}{dx^2} = -\frac{\omega^2}{v^2} f$$

But this is the familiar differential equation satisfied by a sine or cosine function. Remembering that we have defined $x = 0$ as corresponding to one of the fixed ends of the string, with zero transverse displacement at all times, we know that an acceptable solution must be of the form

$$f(x) = A \sin \left(\frac{\omega x}{v} \right) \tag{6-6}$$

But we have the further boundary condition that the displacement is always zero at $x = L$. Hence we must also have

$$A \sin \left(\frac{\omega L}{v} \right) = 0$$

Therefore,

$$\frac{\omega L}{v} = n\pi \tag{6-7}$$

where n is any (positive) integer.

It will be convenient to introduce the number of cycles per unit time, ν, equal to $\omega/2\pi$. The frequencies of the permitted stationary vibrations are thus given by

$$\nu_n = \frac{nv}{2L} = \frac{n}{2L} \left(\frac{T}{\mu} \right)^{1/2} \tag{6-8}$$

165   The free vibrations of stretched strings

where n, according to this calculation, may be 1, 2, 3, ... to infinity.[1]

A vivid way of describing the shape of the string at any instant, in any particular mode n, is obtained by recognizing that the total length of the string must exactly accommodate an integral number of half-sine curves, as implied by Eq. (6–7). We can therefore define a *wavelength*, λ_n, associated with the mode n, such that

$$\lambda_n = \frac{2L}{n} \tag{6–9}$$

Then we can put

$$\frac{\omega}{v} = \frac{n\pi}{L} = \frac{2\pi}{\lambda_n}$$

Hence, from Eq. (6–6), the shape of the string in mode n is characterized by the following equation:

$$f_n(x) = A_n \sin\left(\frac{2\pi x}{\lambda_n}\right) = A_n \sin\left(\frac{n\pi x}{L}\right) \tag{6–10}$$

and the complete description of the motion of the string is thus as follows:

$$y_n(x, t) = A_n \sin\frac{2\pi x}{\lambda_n} \cos \omega_n t \tag{6–11}$$

where

$$\omega_n = \frac{n\pi}{L}\left(\frac{T}{\mu}\right)^{1/2} = n\omega_1$$

Since all the possible frequencies of a given stretched string are, according to the above analysis, simply integral multiples of the lowest possible frequency, ω_1, a particular interest attaches to this basic mode—the *fundamental*. It is the frequency of the fundamental that defines what we recognize as the characteristic pitch of a vibrating string, and which therefore defines for us the tension required to obtain a certain note from a string of given mass and length.

Example. The E string of a violin is to be tuned to a frequency of 640 Hz. Its length and mass (from the bridge to the end) are 33 cm and 0.125 g, respectively. What tension is required?

---

[1]The essential *form* of this functional dependence of v on L, T, and μ was discovered by Galileo.

From Eq. (6–8) we have

$$\nu_1 = \frac{1}{2L}\left(\frac{T}{\mu}\right)^{1/2}$$

and we shall put $\mu = m/L$, where m is the total mass. This then gives us

$$T = 4mL\nu_1{}^2$$
$$= 4(1.25 \times 10^{-4})(0.33)(6.4 \times 10^2)^2 \quad \text{(MKS)}$$
$$\approx 68 \text{ N}$$

This is therefore a pull of about 15 lb.[1] (The total pull of all four strings in an actual violin is about 50 lb.)

THE SUPERPOSITION OF MODES ON A STRING

In a stringed instrument such as a piano, the string is struck once at some chosen point. At the moment of impact, and for a brief instant thereafter, the string is sharply pushed aside near this point, and its shape is nothing like a sine curve. Shortly thereafter, however, it settles down to a motion which is a simple super-position of the fundamental and a few of its lowest harmonics. It is a physically very important fact that these vibrations can occur simultaneously and to all intents independently of one another. It can happen because the properties of the system are such that the basic dynamical equation (6–2) is *linear*—i.e., only the first power of the displacement y occurs anywhere in it. If various individual solutions of this equation, corresponding to the various individual harmonics, are denoted y_1, y_2, y_3, etc., then the sum of these also satisfies the basic equation, and the motion thereby described can always be considered as resolvable into these individual components. Figure 6–3 shows some examples of such compound or superposed vibrations. Their mutual independence can be demonstrated by suddenly stopping the transverse motion of the string at a point that is a node for some harmonics but not for others. Those component vibrations for which the point is a node will continue unaffected; the others will be quenched. Thus, for example, if a piano string has been set sounding loudly by striking the key, which is kept held down, and the string is then touched one third the way along its length, all component vibrations are stopped except the third, sixth, etc., multiples of the fundamental frequency.

[1]Ten newtons $\approx$ 2.2 lb.

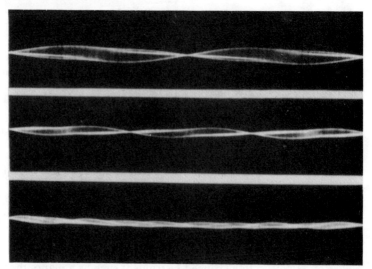

Fig. 6–3 Compound vibrations of a string, made up of combinations of simple modes. (From D. C. Miller, The Science of Musical Sounds, Macmillan, New York, 1922.)

This principle of independence and superposition for the various normal modes of a vibrating system will be seen to have a fundamental importance for the analysis of complicated disturbances; indeed, it is a foundation stone for Fourier analysis. The phenomenon as manifested in a vibrating string was first clearly discussed by Daniel Bernoulli in 1753. Because a real string will not, in practice, be perfectly described by our idealized equations, the independence of the separate modes will not be really complete, although in some circumstances it may be very nearly so.

FORCED HARMONIC VIBRATION OF A STRETCHED STRING

As we have seen above, the free vibrations of a string with both ends rigidly fixed are strictly limited to the fundamental frequency and integral multiples thereof. But now, just as we did in Chapter 4 for a simple harmonic oscillator, we shall consider the response of the string to a periodic driving force. For the purpose of having a simple and well-defined discussion, we shall imagine that the end of the string at $x = L$ remains firmly fixed, but that the end at $x = 0$ is vibrated transversely at some arbitrary angular frequency and with an amplitude B.

Just as in Eq. (6–5), we shall suppose a steady-state solution of the form

$$y(x, t) = f(x) \cos \omega t$$

but now subject to the following conditions:

$$y(0, t) = B \cos \omega t$$
$$y(L, t) = 0$$

The basic equation of motion is still Eq. (6–4), so that $f(x)$ must be a sinusoidal function of x. We therefore put

$$f(x) = A \sin (Kx + \alpha)$$

From Eq. (6–4) we then get $K = \omega/v$, so that

$$f(x) = A \sin \left(\frac{\omega x}{v} + \alpha \right)$$

This is just like Eq. (6–6) except for having an adjustable parameter α. From the boundary condition at $x = L$, we then have

$$\sin \left(\frac{\omega L}{v} + \alpha \right) = 0$$

Therefore,

$$\frac{\omega L}{v} + \alpha = p\pi$$

where p is an integer. From the boundary condition at $x = 0$, we get

$$B = A \sin \alpha$$

Therefore,

$$A = \frac{B}{\sin \left(p\pi - \dfrac{\omega L}{v} \right)} \tag{6–12}$$

The implication of this result is that, for a given amplitude of the forced displacement at the extreme end, the response of the string as a whole will be very large whenever the driving frequency is close to one of the natural frequencies defined by Eq. (6–8). Indeed, according to Eq. (6–12) the driven amplitude would become infinitely large at the exact natural frequencies, and the situation at nearby frequencies would be somewhat as shown in Fig. 6–4. We know, however, that the existence of damping forces will eliminate these unreal infinities, and the actual behavior will simply be to have $A/B \gg 1$ for $\omega \approx \omega_n$.

The important feature of the above result is that we build up a large forced response with a small driving amplitude by having the forcing take place at a point which is close to being a *node* of one of the natural vibrations. Clearly, however, it cannot be a node exactly, because by definition we are imposing motion there. Also, in any real system, the kind of large-amplitude response

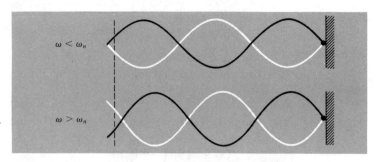

Fig. 6–4 Configurations of string driven just below and just above the natural frequency of a normal mode of vibration.

$\omega < \omega_n$

$\omega > \omega_n$

shown in Fig. 6–4 comes about only after an application of the small periodic driving force over many periods of vibration. There is no magic way of suddenly feeding into the system the large energy represented by the resonant amplitude of vibration. It is very much like the slow growth of a forced, damped oscillator during the transient stage, as indicated in Fig. 4–11(c).

[Anyone who reads this and who is also familiar with the design of pianos may be somewhat puzzled by this matter of driving a string at or near a node to get resonant response. For it is the practice to have the piano hammer strike the string about one seventh of the total length along it. And the purpose and effect of this is to *suppress* the unpleasant-sounding seventh harmonic, not to encourage it, as the above analysis might imply. The point is that any nonzero displacement at a point which is precisely the node of a certain harmonic does not, when applied via a single impulse (as opposed to periodic forcing), represent a means of exciting that particular natural vibration of the system.]

Having seen something of these basic features of the free and forced vibrations of a string, let us now turn to some other systems which exhibit the same kind of behavior.

LONGITUDINAL VIBRATIONS OF A ROD

When the end of a metal rod is struck lengthwise, vibrations of quite high audible frequencies are produced. If the rod is suitably supported—e.g., by a thin clamp at its midpoint—the vibrations persist for quite a time. The Q of the system, especially in the case of the lowest of the possible natural frequencies, is quite high, and may in some cases result in a surprisingly pure tone.

We touched briefly on the properties of this type of system in Chapter 3, in connection with the problem of a body attached to a massive spring. We recognized that the natural frequency of vibration must be proportional to $\sqrt{Y/\rho}$, where Y is the Young's

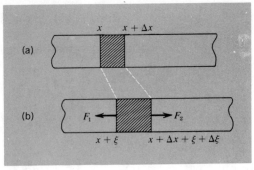

Fig. 6-5 (a) Massive rod. (b) Massive rod after a longitudinal displacement under nonstatic conditions. The shaded section contains the same amount of material as the shaded section in (a).

modulus and ρ is the density. Now we shall go into the situation more carefully.

The problem is actually very much like that of the stretched string, but you are likely to find it a good deal harder to visualize because the displacement is in the same direction as x instead of being transverse to it. We shall use the symbol ξ to designate the displacement, from the equilibrium position, of each particle in the rod that was initially at a distance x from some section that is assumed to be fixed (or at any rate permanently unaccelerated). Then we shall consider the equation of motion of a thin slice of the rod which in the undisturbed state is contained between x and $x + \Delta x$—Fig. 6–5(a).

Then, as indicated in Fig. 6–5(b) (in which, of course, the displacements are much exaggerated), the material shown shaded is bodily shifted and also stretched. It is pulled in opposite directions by the forces F_1 and F_2. Now the magnitude of F_1 depends on the fractional change of interatomic separations *at* x. Similarly, F_2 depends on the fractional change of separation *at* $x + \Delta x$. These forces will in general be slightly different. As a result of the deformation, however, all the material in our slice is in a state of stress, and we can define an average value of this stress in terms of the over-all strain. The length of the slice (originally Δx) has increased by $\Delta \xi$. Therefore,

$$\text{average strain} = \frac{\Delta \xi}{\Delta x}$$

$$\text{average stress} = Y \frac{\Delta \xi}{\Delta x}$$

We now recognize that we can define the stress *at a particular value of* x as being the value of $Y(\partial \xi / \partial x)$ at that point.[1] And then, for a point Δx farther along, we have

[1]Note the partial derivative, because the stress at a given x is also going to vary with t.

$$(\text{stress at } x + \Delta x) = (\text{stress at } x) + \frac{\partial(\text{stress})}{\partial x} \Delta x$$

Thus, if the cross-sectional area of the rod is α, we have

$$F_1 = \alpha Y \frac{\partial \xi}{\partial x}$$

$$F_2 = \alpha Y \frac{\partial \xi}{\partial x} + \alpha Y \frac{\partial^2 \xi}{\partial x^2} \Delta x$$

and so

$$F_2 - F_1 = \alpha Y \frac{\partial^2 \xi}{\partial x^2} \Delta x$$

That ends the conceptually difficult part of the calculation.

We now apply Newton's law to the material lying between x and $x + \Delta x$. If the density is ρ, its mass is $\rho \alpha \Delta x$. Its acceleration is the second time derivative of the displacement, which is just ξ in the limit of vanishingly small Δx [see Fig. 6–5(b)]. Hence we have

$$\alpha Y \frac{\partial^2 \xi}{\partial x^2} \Delta x = \rho \alpha \Delta x \frac{\partial^2 \xi}{\partial t^2}$$

or

$$\frac{\partial^2 \xi}{\partial x^2} = \frac{\rho}{Y} \frac{\partial^2 \xi}{\partial t^2} = \frac{1}{v^2} \frac{\partial^2 \xi}{\partial t^2} \qquad (6\text{–}13)$$

with $v = (Y/\rho)^{1/2}$. This then is really just like Eq. (6–2) for the stretched string, and we can begin looking for solutions of the type

$$\xi(x, t) = f(x) \cos \omega t \qquad (6\text{–}14)$$

There is, however, an important difference of boundary conditions. In most circumstances we shall not have both ends of the rod fixed. It could be arranged, but usually the rod will be clamped either at one end, leaving the other end free, or at the center, leaving both ends free.

We shall just consider the case where one end is fixed. This comes nearest to our earlier, primitive consideration of the oscillation of a massive spring (Chapter 3). Let the fixed end be at $x = 0$, and the free end at $x = L$. We know that Eq. (6–13) implies a sinusoidal variation of ξ with x at any instant, and so we can put

$$f(x) = A \sin\left(\frac{\omega x}{v}\right) \qquad (6\text{–}15)$$

just as in Eq. (6–6).

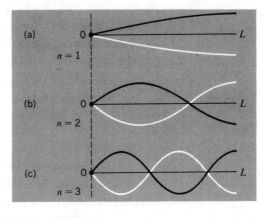

Fig. 6–6 Longitudinal normal modes of massive rod clamped at one end. For clarity the longitudinal displacements are represented as though they were transverse.

The condition at $x = L$ must express the fact that this is a free end. In physical terms this means that the stress there is zero. No adjacent material is pulling on the end of the rod at that point, and conversely there is no adjacent material to be accelerated. Hence, at $x = L$, we have

$$F = \alpha Y \frac{\partial \xi}{\partial x} = 0$$

From Eqs. (6–14) and (6–15), this means

$$\cos\left(\frac{\omega L}{v}\right) = 0$$

or

$$\frac{\omega L}{v} = (n - \tfrac{1}{2})\pi \qquad (6\text{–}16)$$

where n is a positive integer.[1] The natural frequencies of the rod are thus given by

$$\nu_n = \frac{(n - \tfrac{1}{2})v}{2L} = \frac{n - \tfrac{1}{2}}{2L}\left(\frac{Y}{\rho}\right)^{1/2} \qquad (6\text{–}17)$$

Using the condition given by Eq. (6–16), one can see that the length of the rod must accommodate an integral number of *quarter-wavelengths* of sine curves. The three lowest modes are shown schematically in Fig. 6–6, but remember that the displacements are really longitudinal, not transverse.

The lowest mode of such a rod, clamped at one end, has a frequency given by

[1] The use of $(n - \tfrac{1}{2})\pi$ rather than $(n + \tfrac{1}{2})\pi$ in Eq. (6–16) allows us to number the modes 1, 2, 3, etc.

$$\nu_1 = \frac{1}{4L}\left(\frac{Y}{\rho}\right)^{1/2} \tag{6-18}$$

Suppose, for example, we had an aluminum rod, 1 m long. We would have, in this case,

$$Y \approx 6 \times 10^{10}\,\text{kg/m/sec}^2$$
$$\rho \approx 2.7 \times 10^3\,\text{kg/m}^3$$

giving

$$\nu_1 \approx 1200\,\text{Hz}$$

It is interesting to compare our exact result, Eq. (6–18), with what we obtained in Chapter 3 (see discussion on p. 61). Assuming (wrongly) that the stress and strain at any instant during the vibration have the same values along the whole length of the rod, we found the following formula for the frequency of a rod fixed at one end:

$$\text{(wrong)}\quad \nu = \frac{1}{2\pi}\left(\frac{3k}{M}\right)^{1/2} = \frac{\sqrt{3}}{2\pi L}\left(\frac{Y}{\rho}\right)^{1/2}$$

Instead of the coefficient $\frac{1}{4}$ in Eq. (6–18) we would have had $\sqrt{3}/2\pi = 1/3.6$, causing us to overestimate the frequency by about 10%.

THE VIBRATIONS OF AIR COLUMNS

It is clear that a column of air, or other gas, represents a system almost equivalent to a solid rod. Each has its internal elasticity, and the comparison that we began in Chapter 3 can be pressed further in the light of our present discussion.

With an air column it is worth considering all the modes that can be obtained by having either one end or both ends open. An open end represents (approximately, at any rate) a condition of zero pressure change during the oscillation and a place of maximum movement of the air. A closed end, on the other hand, is a place of zero movement and maximum pressure variation. If air is contained' in a tube with one end closed and the other end open, the mode of vibration, and the associated frequency, is defined by one of the situations represented in Fig. 6–7(a), all with a node at one end and an antinode at the other. But it is possible, as shown in Fig. 6–7(b), to get another set of vibrations by leaving both ends of the tube open, hence giving an antinode of displacement at each end. For a tube of a given length, the possible frequencies are then all the integral multiples of the

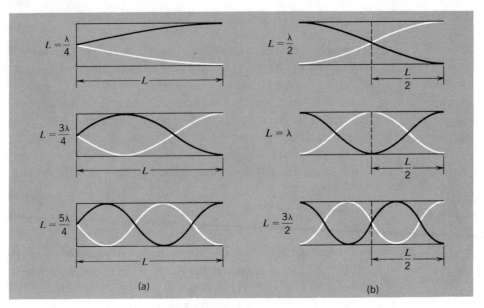

$$L = \frac{\lambda}{4}$$

$$L = \frac{\lambda}{2}$$

$$L$$

$$\frac{L}{2}$$

$$L = \frac{3\lambda}{4}$$

$$L = \lambda$$

$$L$$

$$\frac{L}{2}$$

$$L = \frac{5\lambda}{4}$$

$$L = \frac{3\lambda}{2}$$

$$L$$

$$\frac{L}{2}$$

(a)

(b)

Fig. 6–7 (a) First three normal modes of tube open at one end only. (b) First three normal modes of tube open at both ends.

frequency of the lowest mode with one end closed—the first diagram in Fig. 6–7. The odd multiples all belong to the closed-tube situation, and the even multiples to the open tube. It may be noted that alternate modes in the open-end sequence (those with a node at the center) correspond to the closed-end modes of a column of length $L/2$. It may also be noted that a tube with both ends closed has the same set of natural frequencies as one with both ends open, although it differs from it by an interchange of the positions of nodes and antinodes.[1]

[1]The vibrations of the air columns in actual musical instruments involve many subtleties not even hinted at in this account. The basic *types* of mode are, however, very much as enumerated here. A normal "flue" organ pipe always has an antinode at its mouth (at the bottom end) and may be closed or open at the top. A flute is essentially like an open organ pipe, but in reed instruments (including reed organ pipes and brass instruments in which lips act as reeds) the end with the reed approximates a closed end; the other end is of course open. The reed acts as a driving agent at the resonant frequency of the air column. It feeds in energy at a point approximating to a node of displacement—very much like the string being driven at one end. If you want to read more about this and other fascinating topics in the physics of music, the following books can be recommended as a starting point: Arthur Benade, *Horns, Strings and Harmony*, Doubleday (Science Study Series), New York, 1960; Sir James Jeans, *Science and Music*, Cambridge University Press, New York, 1961; Jess J. Josephs, *The Physics of Musical Sound*, Van Nostrand (Momentum Books), Princeton, N.J., 1967; John Backus, *The Acoustical Foundations of Music*, W. W. Norton, New York, 1969.

175   The vibrations of air columns

The preceding description allows us to enumerate the relative frequencies of an air column, but now let us consider the *absolute* frequency for a gas column of a given length. We are essentially concerned with the correct evaluation of the speed v that appears in the basic differential equation—one just like Eq. (6–13). And this means that we must use an appropriate elastic modulus, K, for the gas, to use in place of the Young's modulus, Y. In Chapter 3 we pointed out that the modulus of a gas was to be defined by the equation

$$K = -V\frac{dp}{dV}$$

and that for the oscillations of a gas the variations of p and V take place under adiabatic conditions—conditions of no heat transfer into or out of the gas—which means that the temperature rises and falls and hence that Boyle's law does not describe the relation between p and V. Now we shall make an explicit calculation.

Suppose that a tube of cross-sectional area A and length l, closed by a piston, contains gas at a pressure p and of density ρ (Fig. 6–8). According to the kinetic theory for an ideal gas, the pressure is given by

$$p = \tfrac{1}{3}\rho v_{\text{rms}}^2 \qquad\qquad (6\text{--}19)$$

where v_{rms}^2 is the mean squared speed of the molecules.[1] If the total mass of gas in the tube is m, we can rewrite Eq. (6–19) as follows:

$$p = \frac{m}{3Al}v_{\text{rms}}^2$$

And this can be written more simply if we introduce the total kinetic energy of translation, E_k, of all the particles:

$$E_k = \tfrac{1}{2}mv_{\text{rms}}^2$$

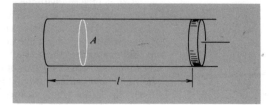

Fig. 6–8   Tube with piston.

[1]The suffix rms stands for root mean square.

Substituting this, we find

$$p = \frac{2}{3A}\frac{E_k}{l} \qquad (6\text{-}20)$$

We shall now consider a movement of the piston, causing a change of the pressure throughout the gas column, and performed in such a way that the work done on the gas by the piston is retained within the gas, thus representing a change in its internal energy. The force needed to cause a compression is essentially equal to pA, so that the work done *on* the gas, in consequence of a change of length, Δl, of the gas column, is given by

$$\Delta W = -pA\,\Delta l$$

and so is positive if Δl is negative. If we assume that this work goes exclusively into increasing the kinetic energy of translation of the molecules, we have

$$\Delta E_k = -pA\,\Delta l \qquad (6\text{-}21)$$

However, the change of length Δl is accompanied by changes of p as well as of E_k; from Eq. (6–20) we have, by differentiation,

$$\Delta p = \frac{2}{3A}\left(\frac{1}{l}\Delta E_k - \frac{\Delta l}{l^2}E_k\right)$$

Therefore,

$$\Delta p = \frac{2}{3Al}\Delta E_k - \frac{\Delta l}{l}\left(\frac{2}{3A}\frac{E_k}{l}\right)$$

But, with the help of Eqs. (6–20) and (6–21), this becomes

$$\Delta p = \frac{2}{3Al}(-pA\,\Delta l) - \frac{\Delta l}{l}(p)$$

$$= -\tfrac{5}{3}p\frac{\Delta l}{l}$$

Since the cross-sectional area of the gas column is assumed to remain unchanged, the value of $\Delta l/l$ can be equated to the fractional volume change $\Delta V/V$. Hence we have

$$K_{\text{adiabatic}} = -V\frac{\Delta p}{\Delta V} = \tfrac{5}{3}p \qquad (6\text{-}22)$$

This is to be compared to the isothermal elastic modulus, which is just equal to p (see Chapter 3). The speed v defined by this adiabatic value of K is thus given by

$$v = \left(\frac{1.667p}{\rho}\right)^{1/2}$$

177 The elasticity of a gas

Actually this expression works for some gases, but not all—and not for air itself. What are the assumptions that have gone into producing it? They are, first, that the work done on the gas in compression is all used to increase the energy of the gas, rather than going into heat losses to the surrounding material; second, that this energy retained within the gas goes entirely into raising the kinetic energy of translation of the molecules, rather than being used in part to increase the energy of their internal motions.

The first condition appears to be satisfied for acoustic vibrations in all gases. The second condition, however, works only for molecules that really behave like hard billiard balls—which points in particular to the monatomic gases He, Ne, A, etc. For other gases, including air, some of the work done on (or by) the gas results in changes of the internal rotations or vibrations of the molecules. Hence, for a given change of volume, the change of kinetic energy of translation—which determines the pressure, according to Eq. (6–20)—is less than our calculation would imply, and so, as we said in Chapter 3, the elasticity of a gas in adiabatic vibration is expressible as

$$K_{\text{adiabatic}} = \gamma p \tag{6-23}$$

where $1 < \gamma \leq \frac{5}{3}$.

For air the value of γ is found to be close to 1.40, and for air at room temperature and ordinary pressure we have

$$p \approx 1.0 \times 10^5 \text{ N/m}^2$$
$$\rho \approx 1.2 \text{ kg/m}^3$$

Thus, for example, if we had a tube 1 m long, closed at one end, its lowest mode would, by analogy with Eq. (6–18), have a frequency given by

$$\nu_1 = \frac{1}{4L} \left(\frac{\gamma p}{\rho} \right)^{1/2}$$
$$\approx 84 \text{Hz}$$

A COMPLETE SPECTRUM OF NORMAL MODES

In the preceding sections we have discussed the natural vibrations, otherwise called normal modes, of various types of physical system. Setting aside the differences in detail, the systems were assumed to have the following features in common:

1. Each system was taken to be effectively one-dimensional and of limited length.

2. Each system was taken to be continuous and uniform in its structure.

3. Each system was subject to boundary conditions at its ends.

4. Each system was controlled by restoring forces proportional to the displacement from equilibrium.

It followed from these conditions that each system possessed a whole set of distinct modes of vibration, each mode characterized by a mode number n, a frequency ν_n, and a wavelength λ_n—this last being simply related to the length L of the system and to the mode number. It also emerged that the characteristic frequencies varied linearly with n for any given system. Let us now ask some questions about these results.

First, how many different modes can a given system have? If one accepts our treatment of the problem at its face value, the number of distinct modes is infinite—although discrete. Is this true? Not quite. If you have followed through the discussion of the last chapter, you will have learned that a line of N interacting particles has a total of N different normal modes of vibration of a given type (e.g., purely transverse or purely longitudinal). For example, a rod 1 m long should be thought of as made up of lines of atoms separated from one another by distances of the order of 1 Å. Thus it would have only about 10^{10} normal modes, instead of infinitely many. But of course 10^{10} is a monstrous number—almost infinite for most physical considerations. The main point, though, is that we can envisage a *complete set*, or *spectrum*, of all the possible normal modes of a given system, and can be sure that, by allowing the mode number n to take on all its possible values, from 1 to N or from 1 to ∞, we have enumerated them all.

Second, what are the permitted wavelengths of the stationary vibrations on a uniform one-dimensional system of a given length L? This depends on the boundary conditions. We shall be devoting special attention to the case in which the displacement is zero at both ends. In that case, as we have seen, the wavelengths λ_n are given by the particularly simple relation—Eq. (6–9):

$$\lambda_n = \frac{2L}{n}$$

This is a purely geometrical result, in the sense that it depends only on the *form* of the equations of motion, and has nothing to do with such quantities as the elasticity and density of the medium. It is good for any value of n.

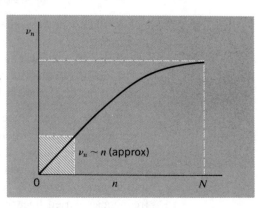

Fig. 6–9 Variation of the mode frequency with mode number up to maximum possible frequency for a one-dimensional system. In the shaded area the relation between ν_n and n is very nearly linear.

Third, what about the *frequencies* of the normal modes? Does the linearity of ν_n versus n hold for very high modes, in the way that it seems to do, with great exactitude, for those first few dozen harmonics that alone are of importance in acoustic phenomena? It will be remembered that, for a system with both ends fixed, we had the particularly simple result—Eq. (6–8):

$$\nu_n = \frac{nv}{2L}$$

where v is a speed that appears to be uniquely defined by the inertial and restoring properties of the medium. But this simple proportionality of ν_n to n is *not* generally true. For a simple, one-dimensional system, the frequency increases less and less rapidly with mode number until a limiting frequency is reached at the highest possible mode number, N, as shown in Fig. 6–9.[1] The system is not capable of vibrating at any higher frequency than this. In the type of one-dimensional system we have described, this phenomenon is tied very directly to the fact that the number of particles participating in the motion is not infinitely large. In fact, however, the proportionality of ν_n to n is a rather special result, and ceases to apply with any generality when we go to two- and three-dimensional systems.[2] Nevertheless, the important feature remains—that a given one-dimensional system, with specified boundary conditions, has a denumerable set (even if infinite) of characteristic natural modes of vibration.

[1] For further discussion, refer back to Chapter 5.

[2] Even in a one-dimensional system that is treated as being continuous in structure, the relation between the mode frequencies becomes quite different if the system is not uniform—e.g., a stretched string whose thickness increases continuously from one end to the other, or (as shown in Fig. 5–1) a uniform chain hanging vertically, in which the tension decreases steadily from the top downward.

Let us end this general discussion of the normal modes by drawing attention to two features, already referred to, that are of especially great importance:

1. The *boundary conditions*, as applied in this case at the two ends of the one-dimensional system, play a decisive role in determining the character of the normal modes.

2. Given the *linearity* of our basic equations of motion, any or all of the normal modes of vibration can coexist with arbitrary relative values of amplitude and phase.

NORMAL MODES OF A TWO–DIMENSIONAL SYSTEM

We shall turn our attention now to a brief consideration of the normal modes of systems that are essentially two-dimensional, such as a stretched elastic sheet or a thin metal plate. As with the one-dimensional systems, the specification of boundary conditions —now, primarily, around the edges—limits the permissible motions to a few particular classes: the normal modes that are consistent with the stated boundary conditions. The precise character of the normal modes may be very beautifully indicative of any symmetries that a given physical system possesses.

The simplest case to consider is that of a rectangular vibrating membrane. By analogy with the one-dimensional case of a vibrating string, a rectangular membrane with a fixed outer boundary has normal modes describable by sines and cosines as follows:

$$z(x, y, t) = C_{n_1 n_2} \sin\left(\frac{n_1 \pi x}{L_x}\right) \sin\left(\frac{n_2 \pi y}{L_y}\right) \cos \omega_{12} t$$

where the normal mode frequencies are

$$\omega_{12} = \left(\frac{S}{\sigma}\right)^{1/2} \left[\left(\frac{n_1 \pi}{L_x}\right)^2 + \left(\frac{n_2 \pi}{L_y}\right)^2\right]^{1/2}$$

(6–24)

Here

S = force/unit length (surface tension)

σ = mass/unit area

$n_1, n_2 = 1, 2, 3, \ldots$

L_x, L_y = lengths of the sides of the membrane.

This equation may look a little formidable, but its structure can be clearly recognized if one compares it to the equation for the normal modes of a stretched string:

$$y_n(x, t) = A_n \sin\left(\frac{n\pi x}{L}\right) \cos \omega_n t$$

where

$$\omega_n = \frac{n\pi}{L}\left(\frac{T}{\mu}\right)^{1/2}$$

The product of sine functions in Eq. (6–24) guarantees that the displacement z is always zero along the boundaries, and the value of ω is characterized by two integers instead of one. The complete membrane can be imagined, if one so desires, as a whole set of strings stretched parallel to one another, along (let us say) the x direction, and all satisfying the same wavelength condition with respect to that direction. But these strings must also be considered as cross-connected along lines parallel to the y direction, satisfying some other wavelength condition between $y = 0$ and $y = L_y$.

The actual dynamics can be constructed by considering a small rectangular patch of the membrane at some arbitrary x and y [see Fig. 6–10(a)]. The surface tension S acts as a certain force per unit length exerted in the surface perpendicular to any line considered. Thus if our patch has edges of length Δx and Δy, the force in the xz plane at each end is $S \Delta y$, and the force in the yz plane is $S \Delta x$. The mass of the patch is $\sigma \Delta x \Delta y$. Considering now the side view in the xz plane, as shown in Fig. 6–10(b), the situation is just like that shown in Fig. 6–2. The transverse force due to the curvature of the membrane in the xz plane is given by

$$S \Delta y \Delta\theta_x = S \frac{\partial^2 z}{\partial x^2} \Delta x \Delta y$$

Fig. 6–10 (a) Force diagram for a small area $\Delta x \Delta y$ of an elastic membrane. (b) Cross section of force diagram in the xz plane.

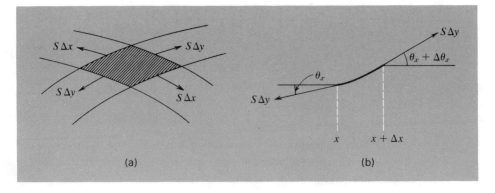

(a)

(b)

But the membrane is curved in the yz plane, too, and the force due to this is given by

$$S \, \Delta x \, \Delta \theta_y = S \frac{\partial^2 z}{\partial y^2} \Delta x \, \Delta y$$

The total force is the sum of these, and the mass being accelerated is $\sigma \, \Delta x \, \Delta y$. Hence the equation of motion ($ma = F$) becomes

$$\sigma \Delta x \, \Delta y \frac{\partial^2 z}{\partial t^2} = S \left(\frac{\partial^2 z}{\partial x^2} + \frac{\partial^2 z}{\partial y^2} \right) \Delta x \, \Delta y$$

or

$$\frac{\partial^2 z}{\partial x^2} + \frac{\partial^2 z}{\partial y^2} = \frac{\sigma}{S} \frac{\partial^2 z}{\partial t^2} \tag{6-25}$$

This equation of motion is a direct extension of Eq. (6–2). By recognizing that there may exist stationary vibrations of the form

$$z(x, y, t) = f(x) g(y) \cos \omega t$$

we come quite straightforwardly to the exact expressions for f, g, and ω that are given in Eq. (6–24).

As we said in the last section, the mode frequencies for a two-dimensional system do not in general exhibit simple numerical relationships—not even in the simplest possible geometry, a square. On the other hand, the points of zero displacement at all t are connected by nodal lines—straight lines parallel to the sides of the rectangle—which correspond in a very simple way to the geometrical conditions imposed at the boundaries.

In Fig. 6–11 we show the few lowest modes of a rectangle that is almost but not quite a square. Its sides are taken to be of lengths $1.05L$ and $0.95L$, giving an area almost exactly equal to L^2. It is convenient to express the possible frequencies as multiples of a frequency ω_1 defined by

$$\omega_1 = \frac{\pi}{L} \left(\frac{S}{\sigma} \right)^{1/2} \qquad \left[\nu_1 = \frac{1}{2L} \left(\frac{S}{\sigma} \right)^{1/2} \right]$$

This is the lowest frequency with which a square membrane of side L could vibrate if it were fixed along two opposite edges and were free along the other two edges. From Eq. (6–24), the normal mode frequencies of our rectangle are given by

$$\omega_{12} = \omega_1 \left[\left(\frac{n_1}{1.05} \right)^2 + \left(\frac{n_2}{0.95} \right)^2 \right]^{1/2}$$

On Fig. 6–11 are shown the nodal lines (dashed) with shading to

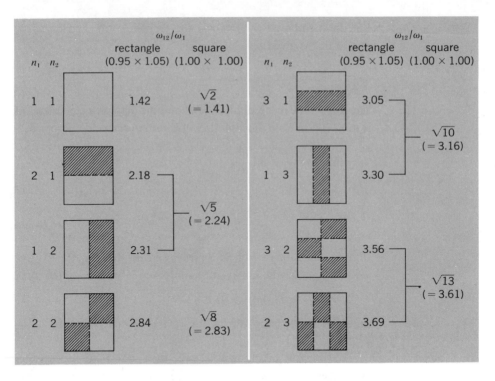

Fig. 6–11 Normal modes of plane rectangular surface compared to those of a square of the same area. Shaded areas and clear areas have displacements of opposite sign perpendicular to the plane of the diagram and passing through zero at the nodal lines.

show which portions of the membrane have displacements in the same direction at any instant.

The normal mode frequencies for a perfectly square membrane would be $\sqrt{2}\,\omega_1$, $\sqrt{5}\,\omega_1$, $\sqrt{8}\,\omega_1$, etc. We have deliberately chosen something that is almost but not quite square so as to draw attention to an interesting and important feature. We notice a tendency for the modes of our rectangle to classify themselves in pairs in which the values of n_1 and n_2 (if different) are interchanged. The frequencies of these paired modes are quite similar and they bracket the frequency that both of these modes would have in a perfectly square membrane. The limiting case—the perfect square—is what is called *degenerate*; a single frequency may correspond to two geometrically distinct patterns of vibration, and the number of normal modes is greater than the number of distinct frequencies. There are other circumstances, too, in which the vibrations of a rectangular membrane may be degener-

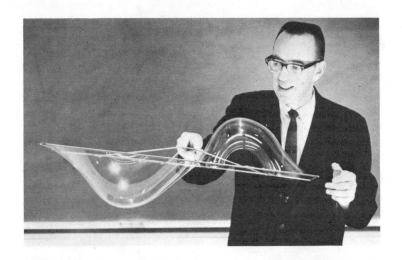

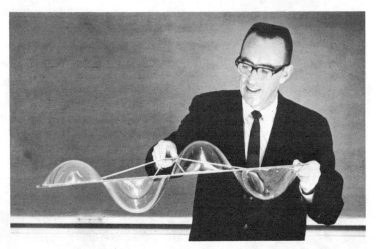

Fig. 6–12 Normal modes of soap film. (Demonstrated by Prof. A. M. Hudson, using a specially strong soap film solution compounded of detergent, glycerin, and a little sugar.)

ate. If, for example, the ratio L_x/L_y is expressible as a ratio of integers, at least two different sets of the numbers (n_1, n_2) may be found which lead to the same value of the frequency. This phenomenon of degeneracy is important, not only in classical mechanics, but also in atomic and nuclear systems. As an example, in the original Kepler-like model of the hydrogen atom, developed by Bohr and Sommerfeld, the electron is pictured as traveling around the proton in certain "allowed" orbits—not just the circular ones that Bohr originally proposed, but a variety of ellipses corresponding to different values of the orbital angular momentum. Many of these distinct orbits correspond in the simplest form of the theory to the same total energy for the electron, but the fact that they are all different is important when

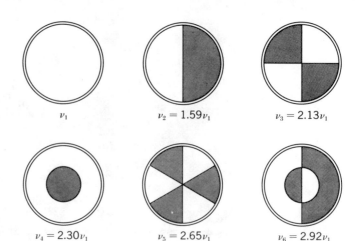

ν_1 $\nu_2 = 1.59\nu_1$ $\nu_3 = 2.13\nu_1$

Fig. 6–13 Normal modes of disk. Shaded area and clear areas have displacements of opposite sign, passing through zero at the nodal lines.

$\nu_4 = 2.30\nu_1$ $\nu_5 = 2.65\nu_1$ $\nu_6 = 2.92\nu_1$

it comes to counting the number of distinguishable states available to an electron in an atom.

The vibrations of soap films, formed on a wire frame that defines a rigid boundary, provide a vivid demonstration of normal modes. Figure 6–12 shows two of the modes of a rectangular membrane as obtained in this way.

Another very important class of vibrations on two-dimensional systems is obtained when the boundary is circular. If again

Fig. 6–14 Displacement maxima of modes of soap films. (Photos by Ludwig Bergmann, supplied by Prof. U. Ingard, M.I.T.)

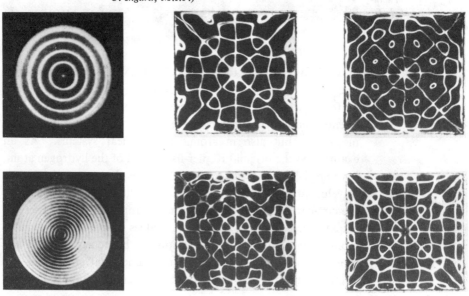

we take this boundary to be fixed, the normal modes express the symmetry of the arrangement by being concentric circles and diametral lines. A rich variety of vibrations is possible; Fig. 6–13 illustrates the lowest six modes of such a system.

Fig. 6–15   Chladni figures showing nodal lines. (From Mary Waller, Chladni Figures: A Study in Symmetry, *Bell, London, 1961.)*

More complex modes of rectangular and circular systems can be excited in soap films by driving them from a nearby loud-speaker which is emitting the appropriate frequency. Figure 6–14 shows some examples; in this photograph the white lines are reflections from the displacement maxima, not the nodal lines.

E. F. Chladni (1756–1827) devised a method for making visible the vibrations of a metal plate clamped at one point or supported at three or more points. Fine sand sprinkled on the plate comes to rest along the nodal lines where there is no motion. The plate may be excited by stroking with a violin bow or by holding a small piece of "dry ice" against the plate. Touching a finger at some point will prevent all oscillations except those for which a nodal line passes through the point touched. Figure 6–15 illustrates some particularly beautiful Chladni figures obtained by Mary Waller.

NORMAL MODES OF A THREE–DIMENSIONAL SYSTEM

A solid block of any material always has some degree of elasticity, and in consequence has a spectrum of normal modes of vibration. This will be true even if—just like the strings and membranes we have been discussing—its boundaries are imagined to be held fixed. For example, a jellylike material that completely fills a more or less rigid container can be felt to be vibrating in a complex way if the container is given a sudden blow.

In the case of one-dimensional and two-dimensional systems, we have been able to discuss and display the characteristic modes of transverse oscillation in a rather vivid manner. When we come to three-dimensional systems we do not any longer have a spare dimension, as it were, along which the displacement may be seen to take place. We shall just content ourselves, therefore, with pointing out that one can set up, for three dimensions, a differential equation of motion that is in strict analogy to the equations we have previously set up for one and two dimensions. The equation will be of the form

$$\frac{\partial^2 \Psi}{\partial x^2} + \frac{\partial^2 \Psi}{\partial y^2} + \frac{\partial^2 \Psi}{\partial z^2} = \frac{1}{v^2}\frac{\partial^2 \Psi}{\partial t^2} \tag{6–26}$$

where v is some characteristic speed—e.g., the speed defined by the value of $\sqrt{K/\rho}$, where K is the appropriate bulk modulus of elasticity. The scalar quantity Ψ might then be the magnitude of the pressure at any given position and time. In discussing the normal vibrations of a rod or an air column, we were in effect using a one-dimensional reduction of this equation. The medium in those cases was certainly three-dimensional, but we chose to confine our attention to vibrations describable in terms of one position coordinate only.

We recognize that boundary conditions must now be specified on all the external *surfaces* of the system. For a rectangular block, fixed over its whole boundary, we can imagine a set of normal modes very much like those of a rectangular membrane. But now the nodal points lie on a set of surfaces, and each normal vibration must now be characterized by a set of three integers, instead of two (membrane) or one (string). Further than this, however, we shall not attempt to go. Instead, we shall return now to the study of one-dimensional problems and the coexistence of a number of normal modes in such a system.

FOURIER ANALYSIS

Suppose we have a string of length L fixed at its two ends. Then, as we have seen, it should be able (subject to certain assumptions about the dynamics) to vibrate in any of an infinite number of normal modes. Allowing for the necessary freedom of choice of both amplitude and phase of a given mode, we shall put

$$y_n(x, t) = A_n \sin\left(\frac{n\pi x}{L}\right) \cos(\omega_n t - \delta_n) \tag{6-27}$$

Furthermore, we can imagine that *all* these modes are permitted to be present, so that the motion of the string is completely specified by the following equation:

$$y(x, t) = \sum_{n=1}^{\infty} A_n \sin\left(\frac{n\pi x}{L}\right) \cos(\omega_n t - \delta_n) \tag{6-28}$$

The actual motion of the string may of course be very hard to visualize—but as long as the physical assumptions leading to Eq. (6–27) are justified, we can assume that an arbitrary synthesis of this type is possible.

Imagine now that a flash photograph is made of the oscillating system. This will show its configuration at some specific time t_0. The quantities $\cos(\omega_n t_0 - \delta_n)$ can then be treated just as a set of fixed numbers, and the displacement of the string at any designated value of x can be written as follows:

$$y(x) = \sum_{n=1}^{\infty} B_n \sin\left(\frac{n\pi x}{L}\right)$$

where (6-29)

$$B_n = A_n \cos(\omega_n t_0 - \delta_n)$$

We now make the following assertion: *It is possible to take any form of profile of the string, described by y as a function of x between x = 0 and x = L (subject to the conditions y = 0 at x = 0 and x = L) and analyze it into an infinite series of sine functions as given in Eq. (6-29).*

There may seem to be a large measure of arbitrariness about the above statement. This arbitrariness disappears, however, if one considers the continuous string as the limit, for $N \rightarrow \infty$, of a row of N connected particles. This is where the insights provided by the discussions of Chapter 5 come to our aid. We could see clearly, for a finite number of particles, how there were precisely N normal modes. The description of each mode involved two adjustable constants—amplitude and phase. *Any* motion of the N particles, under the influence of their mutual interactions, was then describable in terms of a superposition of the normal modes. And the existence of a total of $2N$ adjustable constants allowed us to assign arbitrary values of initial displacement and velocity to every particle. Our present statement is the logical consequence of applying this result to an arbitrarily large number of connected particles.

There is, of course, no actual physical system in which the number of particles is infinite. Thus, in going to this limit, we are, in fact, translating our problem from the world of physics into the world of mathematics. And Eq. (6-29)—a remarkably simple statement—is the basis of one of the most powerful techniques in all of mathematical physics—that of Fourier analysis. The great French mathematician Lagrange (1736-1813), who made mechanics his special province, developed the theory of the vibrating string in just the way that we have chosen to follow, and as long ago as 1759 he came to the verge of enunciating the result expressed in Eq. (6-29). But it was another French mathematician, J. B. Fourier, who (in 1807) was the first to assert that indeed a completely arbitrary function could be described over a given interval by such a series. It is, on the face of things, an extraordinarily unlikely result; it goes against common sense, and yet it is true. We shall shortly consider a specific example of its application, but first let us point to another result that is contained in our dynamical solution for a vibrating system.

Consider the general transverse motion of the continuous

string, as given by Eq. (6–28). According to our original calculations on the continuous string, as developed early in this chapter, the frequencies ω_n are integral multiples of a fundamental frequency ω_1—see Eq. (6–11) and the preceding analysis. If we now fix attention on a particular value of x, we can write $A_n \sin(n\pi x/L)$ as a constant coefficient C_n, and thus have

$$y(t) = \sum_{n=1}^{\infty} C_n \cos(\omega_n t - \delta_n) \qquad \text{where } \omega_n = n\omega_1 \qquad (6\text{–}30)$$

And what *this* states is that any possible motion of any point on the string is periodic in the time $2\pi/\omega_1$, where ω_1 is the frequency of the lowest mode, and further that this periodic motion can be written as a combination, with suitable amplitudes and phases, of pure sinusoidal vibrations comprising all possible harmonics of ω_1. This then is a Fourier analysis in time, rather than in space. You may notice that the expansions expressed by Eqs. (6–29) and (6–30) are of slightly different form. Not only is one made up of sines and the other of cosines, but also, if we cover the whole interval of the variables, we see that $n\pi x/L$ changes by an integral multiple of π, whereas $n\omega_1 t$ changes by an integral multiple of 2π. However, as long as our interest is only in representing the function within the designated range, and not outside it, too, the difference need not concern us.[1]

FOURIER ANALYSIS IN ACTION

To put the Fourier analysis into practice, we must be able to determine the coefficients of the component sine or cosine functions. The process of doing this is called harmonic analysis, and the properties of sine and cosine functions make it a quite simple affair.

Consider the expansion for $y(x)$, as given by Eq. (6–29),

$$y(x) = \sum_{n=1}^{\infty} B_n \sin\left(\frac{n\pi x}{L}\right)$$

[1]Actually, over the range $0 < \omega_1 t < \pi$, an arbitrary function $y(t)$ can be fitted by expressions even simpler than Eq. (6–30) and in strict analogy to Eq. (6–29). It can be described in terms of cosines only, or of sines only, as follows:

cosines only: $y(t) = \sum C_n \cos n\omega_1 t$
sines only: $y(t) = \sum D_n \sin n\omega_1 t$

Because the cosine representation is an even function of $\omega_1 t$ and the sine representation is an odd function, these behave quite differently in the range $\pi < \omega_1 t < 2\pi$.

Suppose we want the amplitude associated with a particular value of n—say n_1. To find it we multiply both sides of the equation by $\sin(n_1\pi x/L)$ and integrate with respect to x over the range from zero to L:

$$\int_0^L y(x) \sin\left(\frac{n_1\pi x}{L}\right) dx = \sum_{n=1}^\infty B_n \int_0^L \sin\left(\frac{n\pi x}{L}\right) \sin\left(\frac{n_1\pi x}{L}\right) dx$$

(6–31)

On the right we still appear to have an infinite series of terms. But now consider the properties of an integral whose integrand is a product of sines. Given any two angles, θ and φ, we have

$$\cos(\theta - \varphi) = \cos\theta \cos\varphi + \sin\theta \sin\varphi$$
$$\cos(\theta + \varphi) = \cos\theta \cos\varphi - \sin\theta \sin\varphi$$

Therefore,

$$\sin\theta \sin\varphi = \tfrac{1}{2}[\cos(\theta - \varphi) - \cos(\theta + \varphi)]$$

Hence we can put

$$\sin\left(\frac{n\pi x}{L}\right) \sin\left(\frac{n_1\pi x}{L}\right) = \frac{1}{2}\left\{\cos\left[\frac{(n - n_1)\pi x}{L}\right] - \cos\left[\frac{(n + n_1)\pi x}{L}\right]\right\}$$

Therefore,

$$\int \sin\left(\frac{n\pi x}{L}\right) \sin\left(\frac{n_1\pi x}{L}\right) dx = \frac{L}{2\pi(n - n_1)} \sin\left[\frac{(n - n_1)\pi x}{L}\right] - \frac{L}{2\pi(n + n_1)} \sin\left[\frac{(n + n_1)\pi x}{L}\right]$$

If we insert the limits $x = 0$, $x = L$, the values of $\sin(n \pm n_1)\pi x/L$ are all zero. Thus at first sight it would appear that we had got rid of the right-hand side of Eq. (6–31) altogether. But then we notice that the quantity $(n - n_1)$ appears in the *denominator* of one of the integrals. Thus if $n = n_1$, we have one integral of the form $0/0$. And it at once turns out that, although all other terms are zero, this one is not. For if $n = n_1$, the integral to be evaluated is the following:

$$\int_0^L \sin^2\left(\frac{n_1\pi x}{L}\right) dx = \tfrac{1}{2}\int_0^L \left[1 - \cos\left(\frac{2n_1\pi x}{L}\right)\right] dx$$

The cosine term contributes nothing between the given limits, but the other part gives us $L/2$. Thus we arrive at the following identity:

192   Normal modes of continuous systems

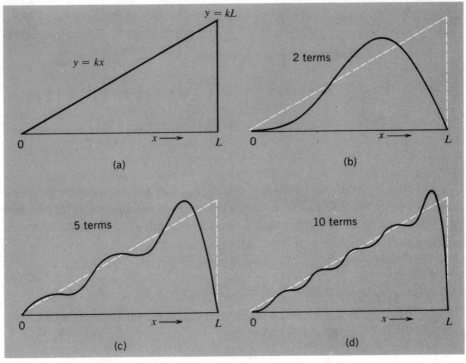

Fig. 6–16 (a) Function to be analyzed—a triangular
sawtooth. (b) (c), (d) Partial Fourier syntheses, using
2, 5 and 10 terms respectively.

$$\int_0^L y(x) \sin\left(\frac{n\pi x}{L}\right) dx = \frac{L}{2} B_n$$

i.e.,

$$B_n = \frac{2}{L} \int_0^L y(x) \sin\left(\frac{n\pi x}{L}\right) dx \qquad (6\text{--}32)$$

This equation determines for us the amplitude B_n associated with
any given value of n in the harmonic analysis of $y(x)$.

If $y(x)$ is a purely empirical curve, the evaluation of the
Fourier coefficients B_n is a matter for computers or graphical
integration. But if the form of $y(x)$ can be described by an exact
analytic function, we can obtain a general formula for all the
B_n's. To illustrate the procedure, let us take the profile shown in
Fig. 6–16(a). This is like the shape of a string that is stretched
aside at one extreme end. (Of course, at $x = L$ exactly we must
have $y = 0$; we are assuming, however, that our equation for y
holds good for points arbitrarily close to the end.) The evaluation
of the B_n's then proceeds as follows:

$$B_n = \frac{2}{L} \int_0^L kx \sin\left(\frac{n\pi x}{L}\right) dx$$

Integrating by parts, we find

$$B_n = \frac{2k}{L}\left\{-\frac{L}{n\pi}\left[x\cos\left(\frac{n\pi x}{L}\right)\right]_0^L + \frac{L}{n\pi}\int_0^L \cos\left(\frac{n\pi x}{L}\right) dx\right\}$$

$$= \frac{2k}{n\pi}\left\{-\left[x\cos\left(\frac{n\pi x}{L}\right)\right]_0^L + \frac{L}{n\pi}\left[\sin\left(\frac{n\pi x}{L}\right)\right]_0^L\right\}$$

$$= -\frac{2kL}{\pi}\frac{\cos n\pi}{n}$$

One recognizes that the values of B_n fall into two categories, according to whether n is odd or even, because the value of $\cos n\pi$ alternates between the values $+1$ and -1. We have, in fact,

$$n \text{ odd: } B_n = \frac{2kL}{n\pi}$$

$$n \text{ even: } B_n = -\frac{2kL}{n\pi}$$

If one wishes, however, one can represent both sets by means of the single formula

$$B_n = (-)^{n+1}\frac{2kL}{n\pi}$$

It is now an easy matter to tabulate the various amplitudes (Table 6–1). Thus our description of the triangular profile becomes

$$y(x) = \frac{2kL}{\pi}\left\{\sin\left(\frac{\pi x}{L}\right) - \frac{1}{2}\sin\left(\frac{2\pi x}{L}\right) + \frac{1}{3}\sin\left(\frac{3\pi x}{L}\right)\cdots\right\}$$

TABLE 6–1: VALUES OF B_n/kL

n	B_n/kL	
1	$\frac{2}{\pi}$	$= 0.636$
2	$-\frac{1}{\pi}$	$= -0.318$
3	$\frac{2}{3\pi}$	$= 0.212$
4	$-\frac{1}{2\pi}$	$= -0.159$
5	$\frac{2}{5\pi}$	$= 0.127$

The result of synthesizing various numbers of terms, using the numerical coefficients of which the first five are listed in Table 6-1, is shown in Figures 6–16(b)–(d). By including further terms we can make the fit as good as we have the patience to achieve. And it is quite remarkable that, with so few terms as we have used, one can simulate the general trend of a profile that differs so radically from any sine curve—especially one that departs so far from zero at one end.

The sine curves in terms of which this Fourier analysis is made represent an example of what are called *orthogonal functions*. The description "orthogonal" belonged originally, of course, to geometry. The orthogonality of two sine functions in Fourier analysis is described by the result

$$\int_0^L \sin\left(\frac{n_1\pi x}{L}\right) \sin\left(\frac{n_2\pi x}{L}\right) dx = 0 \qquad \text{for } n_1 \neq n_2 \qquad (6\text{--}33)$$

This may at first sight appear to have no connection with the geometrical condition, but it is not so far removed as one might think. For if we have two vectors, **A** and **B**, the condition that they are orthogonal (perpendicular) to each other is that their scalar product be zero. In terms of their components this can be written

$$A_x B_x + A_y B_y + A_z B_z = 0 \qquad (6\text{--}34)$$

Now if we replaced the continuous integral of Eq. (6–33) by a summation over a very large number, N, of separate terms (as we might do if we were evaluating the integral by numerical methods), a particular value of x could be written as x_p, where

$$x_p = \frac{pL}{N}$$

Thus Eq. (6–33) would be replaced by the following statement:

$$\frac{L}{N} \sum_{p=1}^{N} \sin\left(\frac{n_1\pi p}{N}\right) \sin\left(\frac{n_2\pi p}{N}\right) = 0 \qquad \text{for } n_1 \neq n_2$$

If we write the condition for orthogonality of two ordinary vectors in the form

$$\sum_{p=1}^{3} A_p B_p = 0 \qquad \text{for } \mathbf{A} \perp \mathbf{B}$$

we see that, in a purely formal sense, the difference between the two statements is merely that one of them involves quantities that are completely described by just three components, whereas the other needs N components (and, in the limit, infinitely many). The possibility of analyzing an arbitrary function in terms of a set of orthogonal functions (not necessarily sines or cosines) is one of the most important and widely used techniques in theoretical physics and engineering.

NORMAL MODES AND ORTHOGONAL FUNCTIONS

We shall end with a few remarks that take us back to actual physical systems. We have seen how the characteristic vibrations of a uniform string are (ideally) describable by sinusoidal functions that are orthogonal in the sense just discussed. Each different mode can exist independently of all the others, and one can thus, in principle, change the amplitude associated with a given mode without affecting any of the others. In this sense, the adjective "normal" applied to the individual modes is a true characterization of their mutual independence—quite analogous to the mutual independence of displacements along perpendicular directions.

Dynamically, too, this orthogonality holds. The total energy of a string vibrating in a superposition of its normal modes is just the sum of the energies for the modes individually. If one writes down the expression for the sum of kinetic and potential energies for a small segment of the string at some arbitrary value of x, it consists of two types of terms: (1) terms involving squares of sines or cosines of the same argument, pertaining to a single mode; (2) terms involving products of sines or cosines of different arguments, representing cross terms from different modes. Because of the orthogonality condition, Eq. (6–33), the terms of type 2 all yield zero when the energy is summed over the whole length of the string. Thus the basic modes are indeed dynamically orthogonal to one another in a very complete way.

Our discussion of this independence, or orthogonality, of the normal modes of a system really began with our analysis of the motions of two coupled pendulums in Chapter 5. You may have chosen to accept our suggestion of possibly bypassing Chapter 5 in a first reading. Whether or not you did this, you may well find it helpful to refer back to the beginning of Chapter 5 at this point, and concentrate on following the main line of development from there, so as to be reminded of the common thread that runs through this whole subject.

6–1 A uniform string of length 2.5 m and mass 0.01 kg is placed under a tension 10 N.

(a) What is the frequency of its fundamental mode?

(b) If the string is plucked transversely and is then touched at a point 0.5 m from one end, what frequencies persist?

6–2 A string of length L and total mass M is stretched to a tension T. What are the frequencies of the three lowest normal modes of oscillation of the string for transverse oscillations? Compare these frequencies with the three normal mode frequencies of three masses each of mass $M/3$ spaced at equal intervals on a massless string of tension T and total length L.

6–3 The derivation of free vibrations of a stretched string in the text ignores gravity. Is this omission justified? How would the analysis proceed if gravitational effects were included?

6–4 Show that the analysis in the text for free vibrations of a horizontal string is also valid for a vertical string if $T \gg mg$.

6–5 A stretched string of mass m, length L, and tension T is driven by two sources, one at each end. The sources both have the same frequency ν and amplitude A, but are exactly 180° out of phase with respect to one another. What is the smallest possible value of ω consistent with stationary vibrations of the string?

6–6 A uniform rod is clamped at the center, leaving both ends free.

(a) What are the natural frequencies of the rod in longitudinal vibration?

(b) What is the wavelength of the nth mode?

(c) Where are the nodes for the nth mode?

6–7 Derive the wave equation for vibrations of an air column. Your final result should be

$$\frac{\partial^2 \xi}{\partial x^2} = \frac{\rho}{K} \frac{\partial^2 \xi}{\partial t^2}$$

where ξ is the displacement from the equilibrium position, ρ is the mass density, and K is the elastic modulus.

6–8 Show that for vibrations of an air column:

(a) An open end represents a condition of zero pressure change during the oscillation and hence a place of maximum movement of the air (an antinode).

(b) A closed end is a place of zero movement (a node) and hence maximum pressure variation.

6–9   A room has two opposing walls which are tiled. The remaining walls, floor, and ceiling are lined with sound-absorbent material. The lowest frequency for which the room is acoustically resonant is 50 Hz.

(a) A complex noise occurs in the room which excites *only* the lowest two modes, in such a way that each mode has its maximum amplitude at $t = 0$. Sketch the appearance, for each mode *separately*, of the displacement versus x at $t = 0$, $t = 1/200$ sec, and $t = 1/100$ sec.

(b) It is observed that the maximum displacement of dust particles in the air (which does not necessarily occur at the same time at each position!) at various points between walls is as follows:

x	$L/4$	$L/2$	$3L/4$
ξ_{max}	$+10\mu$	$+10\mu$	-10μ

What are the amplitudes of each of the two separate modes?

6–10 A laser can be made by placing a plasma tube in an optical resonant cavity formed by two highly reflecting flat mirrors, which act like rigid walls for light waves (see figure). The purpose of the plasma tube is to produce light by exciting normal modes of the cavity.

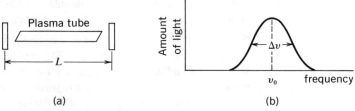

(a)   (b)

(a) What are the normal mode frequencies of the resonant cavity? (Express your answer in terms of the distance L between the mirrors and the speed of light c.)

(b) Suppose that the plasma tube emits light centered at frequency $\nu_0 = 5 \times 10^{14}$ Hz with a spectral width $\Delta\nu$, as shown in the sketch. The value of $\Delta\nu$ is such that all normal modes of the cavity whose frequency is within $\pm 1.0 \times 10^9$ Hz of ν_0 will be excited by the plasma tube.

(1) How many modes will be excited if $L = 1.5$ m?

(2) What is the largest value of L such that only *one* normal mode will be excited (so that the laser will have only one output frequency)? ($c = 3 \times 10^8$ m/sec.)

6–11 (a) Find the total energy of vibration of a string of length L, fixed at both ends, oscillating in its nth characteristic mode with an amplitude A. The tension in the string is T and its total mass is M. (*Hint:* Consider the integrated kinetic energy at the instant when the string is straight so that it has no stored potential energy over and above what it would have when not vibrating at all.)

(b) Calculate the total energy of vibration of the same string if it is vibrating in the following superposition of normal modes:

$$y(x, t) = A_1 \sin\left(\frac{\pi x}{L}\right) \cos \omega_1 t + A_3 \sin\left(\frac{3\pi x}{L}\right) \cos\left(\omega_3 t - \frac{\pi}{4}\right)$$

(You should be able to verify that it is the sum of the energies of the two modes taken separately.)

6–12 A string of length L, which is clamped at both ends and has a tension T, is pulled aside a distance h at its center and released.

(a) What is the energy of the subsequent oscillations?

(b) How often will the shape shown in the figure reappear? (Assume that the tension remains unchanged by the small increase of length caused by the transverse displacements.) [*Hint:* In part (a), consider the work done against the tension in giving the string its initial deformation.]

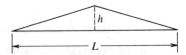

6–13 Consider a uniform cube of side L in which the characteristic wave speed is v. Show that for this system the total number of modes of vibration corresponding to frequencies between ν and $\nu + d\nu$ is

$$4L^3 \nu^2 \, \Delta\nu / \pi^2 v^3 \quad \text{if} \quad \pi v/L \ll \Delta\nu \ll \nu.$$

[*Hint:* Since $\nu L/\pi v = (n_1^2 + n_2^2 + n_3^2)^{1/2}$ consider a cubic lattice of points, letting $x = n_1$, $y = n_2$, $z = n_3$. The number of points in any region of this lattice is thus equal to the volume of that region, and the modes corresponding to a given frequency ν correspond to those points located a distance $r = \nu L/\pi v$ from the origin. The desired result is therefore equal to $4\pi r^2 \, \Delta r$. What would the result be for a square? A rod? How would your answer be altered if a rectangular solid of sides a, b, and c were considered instead of a cube?

6–14 Find the Fourier series for the following functions ($0 \le x \le L$):

(a) $y(x) = Ax(L - x)$.

(b) $y(x) = A \sin(\pi x/L)$.

(c) $y(x) = \begin{cases} A \sin(2\pi x/L), & 0 \le x \le L/2 \\ 0, & L/2 \le x \le L. \end{cases}$

6–15 Find the Fourier series for the motion of a string of length L if

(a) $y(x, 0) = Ax(L - x)$; $(\partial y/\partial t)_{t=0} = 0$.

(b) $y(x, 0) = 0$; $(\partial y/\partial t)_{t=0} = Bx(L - x)$.

Is the ocean composed of water or of waves or of both? Some of my fellow passengers on the Atlantic were emphatically of the opinion that it is composed of waves; but I think the ordinary unprejudiced answer would be that it is composed of water.

SIR ARTHUR EDDINGTON, *New Pathways in Science* (1935)

7

Progressive
waves

WHAT IS A WAVE?

FOR MANY PEOPLE—perhaps for most—the word "wave" conjures up a picture of an ocean, with the rollers sweeping onto the beach from the open sea. If you have stood and watched this phenomenon, you may have felt that for all its grandeur it contains an element of anticlimax. You see the crests racing in, you get a sense of the massive assault by the water on the land—and indeed the waves *can* do great damage, which means that they are carriers of energy—but yet when it is all over, when the wave has reared and broken, the water is scarcely any farther up the beach than it was before. That onward rush was not to any significant extent a bodily motion of the water. The long waves of the open sea (known as the swell) travel fast and far. Waves reaching the California coast have been traced to origins in South Pacific storms more than 7000 miles away, and have traversed this distance at a speed of 40 mph or more. Clearly the sea itself has not traveled in this spectacular way; it has simply played the role of the agent by which a certain effect is transmitted. And here we see the essential feature of what is called wave motion. A condition of some kind is transmitted from one place to another by means of a medium, but the medium itself is not transported. A local effect can be linked to a distant cause, and there is a time lag between cause and effect that depends on the properties of the medium and finds its expression in the velocity of the wave. All material media—solids, liquids, and gases—can carry energy and information by means of waves, and our study of coupled oscillators and normal modes has paved the way for an understanding of this important phenomenon.

Although waves on water are the most familiar type of wave, they are also among the most complicated to analyze in terms of underlying physical processes. We shall, therefore, not have very much to say about them. Instead, we shall turn to our old standby—the stretched string—about which we have learned a good deal that can now be applied to the present discussion.

NORMAL MODES AND TRAVELING WAVES

To set up a particular normal mode of a stretched string, one could make a template of exactly the shape of the string at maximum amplitude in this mode and fit the string to it. Then the sudden removal of this constraint from the string would lead to continuing vibration in this mode alone.

It is much more likely, however, that one would establish the mode by vibrating one end of the string from side to side in simple harmonic motion at the frequency of the mode desired. But what really happens in that case? The stationary vibration does not come into existence immediately. What happens is that a *traveling wave* begins moving along the string. At any instant it is a sinusoidal function of x [Fig. 7–1(a)]. But when the advancing wave reaches the fixed end of the string ($x = L$) there occurs a process of reflection (which we shall consider more carefully in

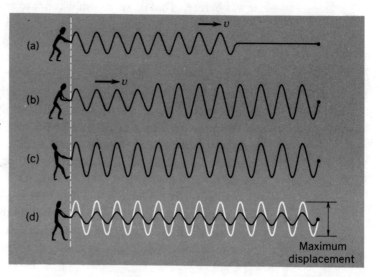

Fig. 7–1 (a) Traveling wave being generated. (b) Traveling wave plus reflected traveling wave. (c) Resultant standing wave (normal mode) at maximum amplitude. (d) Same standing wave as in (c) but at an instant when the displacements are much less than maximum.

Maximum displacement

Chapter 8) and the motion of any point on the string becomes the resultant effect of these two oppositely traveling disturbances [Fig. 7–1(b)]. And after the reflected wave has arrived back at the driven end, there will develop (if the frequency is right in relation to the length and the tension and the mass per unit length of the string) a standing wave which is precisely the normal mode desired [Fig. 7–1(c)]. Thereafter the string continues to vibrate in the manner characteristic of a normal mode; i.e., each point of it continues to vibrate transversely in SHM, and certain nodal points will remain permanently at rest [Fig. 7–1(d)]. Once the normal mode has been established in this way, and the requisite energy fed into it by the driver, the end at $x = 0$ is held stationary.

At this point we can usefully introduce the results of our formal analysis of the normal modes of a stretched string. We found that a continuous string of length L, fixed at both ends, could in principle vibrate in an infinite number of normal modes. These modes are described by the equation

$$y_n(x, t) = A_n \sin\left(\frac{n\pi x}{L}\right) \cos \omega_n t \tag{7-1}$$

where

$$\omega_n = \frac{n\pi}{L}\left(\frac{T}{\mu}\right)^{1/2} \tag{7-2}$$

(T is the tension in the string and μ the mass per unit length.) You will recall that n is an integer, and that if one idealizes to the case of a truly continuous string, then n may run all the way up to infinity.[1] Now let us use a bit of elementary mathematics to cast Eq. (7–1) into a different form. Given any two angles, θ and φ, we have the identity:

$$\sin(\theta + \varphi) + \sin(\theta - \varphi) = 2 \sin \theta \cos \varphi$$

Therefore,

$$\sin \theta \cos \varphi = \tfrac{1}{2}[\sin(\theta + \varphi) + \sin(\theta - \varphi)]$$

Applying this result to Eq. (7–1), we have

$$\sin\left(\frac{n\pi x}{L}\right) \cos \omega_n t = \tfrac{1}{2}\left[\sin\left(\frac{n\pi x}{L} - \omega_n t\right) + \sin\left(\frac{n\pi x}{L} + \omega_n t\right)\right]$$

Hence the nth normal mode for transverse vibrations of the string can be described by the following equation:

[1]Remember, however, that n does have a finite upper limit, and also that Eq. (7–2) for ω_n is strictly only an approximation, which fails when n is large.

203   Normal modes and traveling waves

$$y_n(x, t) = \tfrac{1}{2}A_n \sin\left(\frac{n\pi x}{L} - \omega_n t\right) + \tfrac{1}{2}A_n \sin\left(\frac{n\pi x}{L} + \omega_n t\right)$$

$$\tag{7-3}$$

If in addition we make use of Eq. (7–2) for ω_n, we have

$$y_n(x, t) = \tfrac{1}{2}A_n \sin\left[\frac{n\pi}{L}\left(x - \sqrt{\frac{T}{\mu}}\,t\right)\right]$$

$$+ \tfrac{1}{2}A_n \sin\left[\frac{n\pi}{L}\left(x + \sqrt{\frac{T}{\mu}}\,t\right)\right] \tag{7-4}$$

Finally, as we saw in discussing normal modes in Chapter 6, and as is in any case dimensionally apparent in Eq. (7–4), we can define a characteristic speed v through the equation

$$v = \left(\frac{T}{\mu}\right)^{1/2} \tag{7-5}$$

What we shall now proceed to verify is that Eq. (7–4) is an explicit mathematical description of two traveling waves going in opposite directions.

Suppose we fix attention on the first of the two terms on the right-hand side of Eq. (7–4). It is of the following form:

$$y(x, t) = A \sin\left[\frac{2\pi}{\lambda}(x - vt)\right] \tag{7-6}$$

where $\lambda = 2L/n$. If we imagine first that the time is frozen at some particular instant, the profile of the disturbance is a sine wave with a distance λ between crests (or between any other two successive values of x having the same values of displacement and slope). The quantity λ is, of course, the wavelength of the particular disturbance. Let us now fix attention on any one value of y, corresponding to certain values of x and t, and ask ourselves where we find that same value of y at a slightly later instant, $t + \Delta t$. If the appropriate location is $x + \Delta x$, we must have

$$y(x, t) = y(x + \Delta x, t + \Delta t)$$

Therefore,

$$\sin\left[\frac{2\pi}{\lambda}(x - vt)\right] = \sin\left(\frac{2\pi}{\lambda}[(x + \Delta x) - v(t + \Delta t)]\right)$$

It follows from this that the values of Δx and Δt are related through the equation

$$\Delta x - v\,\Delta t = 0$$

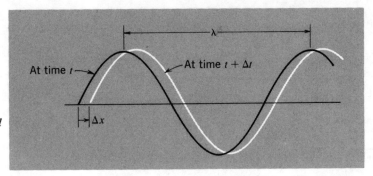

Fig. 7–2 Incremental displacement of wave traveling in the positive x direction.

At time t →

← At time $t + \Delta t$

λ

Δx

i.e.,

$$\frac{\Delta x}{\Delta t} = v$$

What this implies is that, as indicated in Fig. 7–2, the wave as a whole is moving in the positive x direction with speed v.

In an exactly similar way, we can see that the second term in Eq. (7–4) describes a wave of the same wavelength, but traveling in the *negative x* direction with speed v. The standing wave appears to be precisely equivalent, mathematically, to the superposition of these two oppositely moving waves of the same wavelength and amplitude. In saying this, however, we must introduce an important qualification. The curve described by Eq. (7–6), and its counterpart with $(x + vt)$ in the argument of the sine function, represent sine waves of infinite extent—i.e., defined automatically, by the equations, to exist at all x at all t. But the system that we took as our starting point was a string of finite length L, not an infinite one. Thus our new description of the normal mode in terms of traveling waves is not really correct. It is easy, however, to see what lies behind the discrepancy. Figure 7–3 shows several successive stages in the progress of the two oppositely moving waves. Also shown is the result of adding ordinates of the two so as to obtain the resultant displacement as a function of x. At the points A and B, distance L apart, this displacement is zero at all times (as, of course, it was required to be from the original statement of the problem). In between, it varies exactly according to Eq. (7–1). One can say then that, as far as conditions between $x = 0$ and $x = L$ are concerned, the description in terms of infinite wave trains is correct. The fact that there is no continuation of the disturbance outside these limits is a physical condition that was already concealed when we wrote down the equation of a normal mode by means of the single equation

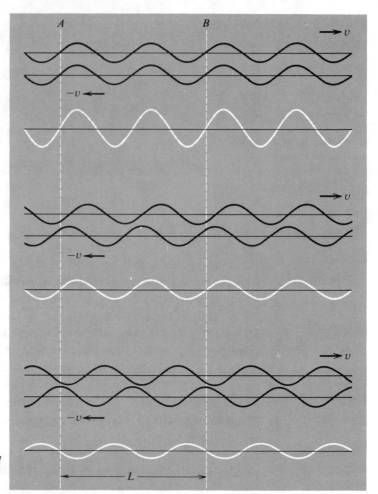

Fig. 7–3 Two ex-
actly similar sinusoi-
dal waves traveling in
opposite directions and
the resultant standing
waves.

$$y_n(x, t) = A_n \sin\left(\frac{n\pi x}{L}\right) \cos \omega_n t$$

because, of course, the function $\sin(n\pi x/L)$ is likewise a function
extending over the whole domain of x. We ought to have been
more careful; the proper description of the vibrating string in
terms of continuous functions of x must be spelled out as follows
for three distinct regions:

$$\left.\begin{array}{l} -\infty \le x < 0: \quad y(x, t) = 0 \\[4pt] 0 \le x \le L: \quad y_n(x, t) = A_n \sin\left(\dfrac{n\pi x}{L}\right) \cos \omega_n t \\[4pt] L < x \le \infty: \quad y(x, t) = 0 \end{array}\right\} \qquad (7\text{–}7)$$

It was important to make the above remarks, because, as

we first remarked in Chapter 1, it is all too easy to forget the limitations that the actual boundary conditions place upon a given physical situation. One is liable, unthinkingly, to allow a mathematical description to wander beyond the limits of its relevance. But having said that, let us now use our imagination to broaden the application of our ideas.

PROGRESSIVE WAVES IN ONE DIRECTION

In the last section we saw how a normal mode of vibration of a stretched string is describable as a combination of two progressive sine waves, identical to one another except for the direction of travel. Why not, then, suppose that on a sufficiently long string it might be possible to set up a sine wave traveling in one direction only? The initiation of such a wave would be carried out exactly as indicated in Fig. 7–1(a), but let us now imagine that the fixed end of the string is very far away—i.e., the total length L of the string is very large compared to the wavelength λ. After a number of cycles of oscillation at $x = 0$, the front end of the disturbance has moved out of the field of view (Fig. 7–4), and the description of all that we see to the right of the plane $x = 0$ is contained in the equation [Eq. (7–6)]

$$y(x, t) = A \sin\left[\frac{2\pi}{\lambda}(x - vt)\right]$$

The generation of this wave comes about as the result of oscillating the left-hand end of the string up and down in SHM of amplitude A and with a frequency ν given by

$$\nu = \frac{v}{\lambda} \quad \text{(or } \omega = 2\pi v/\lambda) \tag{7-8}$$

Explicitly, the equation for y as a function of t at $x = 0$ is

$$y_0(t) = -A \sin\left(\frac{2\pi vt}{\lambda}\right) = -A \sin \omega t$$

The appearance of the *string* at any given time, t_0, is described by

Fig. 7–4   *Generation of traveling wave on a long string.*

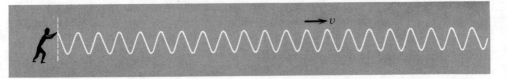

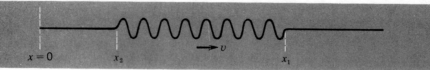

Fig. 7-5   Traveling finite wave train.

$$y(x, t_0) = A \sin\left[\frac{2\pi}{\lambda}(x - vt_0)\right]$$

$$= A \sin\left(\frac{2\pi x}{\lambda} - \varphi_0\right)$$

where φ_0 is a constant angle for the purpose of this instantaneous description of the appearance of the wave. If the end of the string at $x = 0$ were at rest up to $t = t_1$, were vibrated sinusoidally from $t = t_1$ to $t = t_2$, and were kept at rest from t_2 onward, then there would appear on the string a train of sine waves of limited extent, contained at any instant between $x = x_1$ and $x = x_2$, as shown in Fig. 7-5. The front end of the disturbance, farthest away from the end $x = 0$, corresponds to the commencement of the vibration at $t = t_1$, and the rear end to its termination at $t = t_2$. We have, in fact,

$$x_1 - x_2 = v(t_2 - t_1)$$

This is a particular example of a very important result:

The propagation of the wave along the string at the constant speed v is, in effect, a means of translating the variation of displacement with time at a fixed position into a corresponding variation of displacement with position at any designated time.[1]

For any pure sinusoidal disturbance, the wave speed v is definable as the product $\nu\lambda$ [see Eq. (7-8)]. And according to Eq. (7-5), the value of v for waves on a stretched string has the same value, $\sqrt{T/\mu}$, for all wavelengths. This lack of any dependence of v on λ or ν does *not* hold generally true for wave motions. For the time being, however, we shall confine ourselves to situations for which it can be assumed valid.

Let us now set up the differential equation that governs the propagation of a one-dimensional wave as described by Eq. (7-7).

---

[1]There is a concealed subtlety here. As we shall see later, one cannot take it for granted that a sinusoidal vibration of *limited* duration in time will generate a purely sinusoidal wave of limited extent in space. But there will still exist a correspondence between what happens at the source and what appears on the string.

This will be a relation between the partial derivatives of the displacement y with respect to x and t. We have

$$\frac{\partial y}{\partial x} = \frac{2\pi}{\lambda} A \cos\left[\frac{2\pi}{\lambda}(x - vt)\right]$$

$$\frac{\partial y}{\partial t} = -\frac{2\pi v}{\lambda} A \cos\left[\frac{2\pi}{\lambda}(x - vt)\right]$$

Should we then write the differential equation of the wave as

$$\frac{\partial y}{\partial x} = -\frac{1}{v}\frac{\partial y}{\partial t}?$$

There would be nothing to prevent this, but it would cramp our style somewhat, because the above equation applies only to waves traveling in the *positive* x direction. For suppose we take the equation

$$y = A \sin\left[\frac{2\pi}{\lambda}(x + vt)\right]$$

of a wave traveling in the negative x direction. We should then have

$$\frac{\partial y}{\partial x} = \frac{2\pi}{\lambda} A \cos\left[\frac{2\pi}{\lambda}(x + vt)\right]$$

$$\frac{\partial y}{\partial t} = \frac{2\pi v}{\lambda} A \cos\left[\frac{2\pi}{\lambda}(x + vt)\right]$$

and hence

$$\frac{\partial y}{\partial x} = +\frac{1}{v}\frac{\partial y}{\partial t}$$

However, by forming the *second* derivatives, we arrive at a relationship that is true for sine waves of any wavelength traveling in either direction:

$$\frac{\partial^2 y}{\partial x^2} = \frac{1}{v^2}\frac{\partial^2 y}{\partial t^2} \tag{7-9}$$

It comes as no surprise that this is the identical equation of motion from which we started in Chapter 6 [Eq. (6–4)] and which yielded us the normal modes of a stretched string or other continuous one-dimensional system subject to linear restoring forces

WAVE SPEEDS IN SPECIFIC MEDIA

Any system governed by Eq. (7–9) is a system in which sinusoidal waves of any wavelength can travel with the speed v. It may then be a matter of interest to calculate the value of v in any particular case. For example, suppose that a string or wire having $\mu = 0.5$ g/m is stretched with a force of 100 N. For transverse waves on such a string we should have

$$v = \left(\frac{T}{\mu}\right)^{1/2} \approx 450 \text{ m/sec}$$

On the other hand, a rope or length of rubber hose, with a mass per unit length of about 1 kg/m would, if stretched to the same tension, carry waves at only about 10 m/sec—which is actually still quite rapid.

We have developed Eq. (7–9) in terms of transverse waves only; but as we saw in Chapter 6, the longitudinal vibrations of a column of elastic material are governed by an equation of exactly the same form:

$$\frac{\partial^2 \xi}{\partial x^2} = \frac{1}{v^2} \frac{\partial^2 \xi}{\partial t^2} \tag{7–10}$$

This is the basic differential equation for compressional waves traveling along one dimension—waves of a type that can be lumped together under the general title of *sound*, even though only a limited range of their frequencies is detectable by the human ear. It is appropriate at this point to consider the speed of such sound waves in different materials.

1. *Solid bars.* The value of v for waves traveling along the length of a bar or rod is defined by the Young's modulus and the density:

$$v = \left(\frac{Y}{\rho}\right)^{1/2}$$

Table 7–1 shows some data on Young's modulus, density, and the calculated and observed speeds of sound in various materials. It may be seen that speeds of several thousand meters per second

TABLE 7–1:   YOUNG'S MODULI AND SOUND VELOCITIES

Material	$Y, N/m^2$	kg/m^3	$\sqrt{Y/\rho}, m/sec$	$v, m/sec$
Aluminum	6.0×10^{10}	2.7×10^3	4700	5100
Granite	5.0×10^{10}	2.7×10^3	4300	~5000
Lead	$\sim 1.6 \times 10^{10}$	11.4×10^3	1190	1320
Nickel	21.4×10^{10}	8.9×10^3	4900	4970
Pyrex	6.1×10^{10}	2.25×10^3	5200	5500
Silver	7.5×10^{10}	10.4×10^3	2680	2680

are typical, and that the agreement between calculated and observed values is not too bad. It is worth remembering that the Young's modulus is based on static measurements, whereas the propagation of sound depends on the response of the material to rapidly alternating stresses, so exact agreement is not necessarily to be expected. Also, the use of Young's modulus assumes that the material is free to expand or contract sideways (very slightly, of course) as the wave of compression or decompression passes by. But bulk material is not free to do this; the resistance to deformation is in effect increased, and so the calculated speed is raised. The difference is not enormous, however (it is of the order of 15%), and for the purpose of the present discussion we shall not consider it further.

The speed of these elastic waves in solids is notably high. A compressional wave in granite, for example, such as might be generated by an earthquake, has a speed of about 5 km/sec, and would travel about halfway around the earth in the space of 1 hr.

2. *Liquid columns.* A liquid, like a gas, is characterized in its elastic behavior by its bulk modulus, K. Liquids are, in general, far more compressible than solids, without being very much less dense; this means that sound waves travel in liquids more slowly than in solids. The most important case is water. The volume of water is decreased by about 2.3% by application of a pressure of about 500 atm (1 atm $\approx 10^5$ N/m^2). This gives a bulk modulus of about 2.2×10^9 N/m^2, and as $\rho \approx 10^3$ kg/m^3, we have

$$v = \left(\frac{K}{\rho}\right)^{1/2} \approx 1500 \text{ m/sec}$$

This is quite close to the actual figure, and most liquids carry compressional waves at a speed of the order of 1 km/sec.

3. *Gas columns.* We saw in Chapter 6 how the frequencies of vibration of a gas column depend on an adiabatic modulus of elasticity that may differ very significantly from the isothermal modulus. This large difference arises because of the high compressibility of a gas, which means that substantial amounts of work are done on it if the pressure is changed. Although the vibrations in a solid or a liquid may also be adiabatic, the much smaller compressibility means that relatively far less energy can be accepted in this way, and the isothermal and adiabatic elastic moduli are not very different.

In Chapter 6 we pointed out that the adiabatic elasticity modulus of a gas is given by

$$K_{\text{adiabatic}} = \gamma p \qquad (1 < \gamma \le \tfrac{5}{3})$$

so that

$$v = \left(\frac{\gamma p}{\rho}\right)^{1/2} \tag{7-11}$$

For air, $\gamma \approx 1.4$, $\rho \approx 1.2 \text{ kg/m}^3$, and this gives

$$v \approx 340 \text{ m/sec}$$

It is worth giving a little more attention to Eq. (7–11). The general gas equation for a mass m of an effectively ideal gas of molecular weight M is

$$pV = \frac{m}{M} RT$$

where R is the gas constant and T is the absolute temperature. Since the ratio m/V is just the density ρ, Eq. (7–11) would give us

$$v = \left(\frac{\gamma RT}{M}\right)^{1/2} \tag{7-12}$$

The velocity of sound in a gas would thus be expected to be (a) independent of pressure or density, (b) proportional to the square root of the absolute temperature, and (c) inversely proportional to the square root of the molecular weight. Results (a) and (b) are correct for any given gas, at least over a wide range of p or T, and (c) is borne out if we compare various gases of the same molecular type (e.g., all diatomic).

The other particularly interesting feature about Eq. (7–11) comes to light if we recall the simple kinetic theory calculation of the pressure of a gas. This calculation leads to the result [Eq. (6–19), p. 176]:

$$p = \tfrac{1}{3}\rho v_{\text{rms}}^2$$

where v_{rms} is the root-mean-square speed of the molecules. From this result we therefore have

$$v_{\text{rms}} = \left(\frac{3p}{\rho}\right)^{1/2} \tag{7-13}$$

Comparing Eqs. (7–11) and (7–13), we see that the speed of sound in a gas, as given by our calculation, is just about equal to the mean speed of the molecules themselves. As the information that (for example) one end of a gas column has been struck must be carried by the molecules themselves, this approximate equality of sound and molecular speeds (at a few hundred meters per second) makes good sense.

We have seen how it is possible to cause a stretched string to vibrate in a superposition made up of an arbitrary selection of its normal modes. Let us now consider the closely related problem of setting up progressive waves of several different wavelengths on a long string or other such medium. To begin with, let us take the very simple case of two waves of equal amplitude, both traveling along the positive x direction and separately described by equations of the form of Eq. (7–7):

$$y_1 = A \sin\left[\frac{2\pi}{\lambda_1}(x - vt)\right]$$
$$y_2 = A \sin\left[\frac{2\pi}{\lambda_2}(x - vt)\right]$$

(7–14)

Because of what we have learned about the linear superposition of displacements in systems obeying equations like Eq. (7–9), we know that the resultant displacement is just the sum of y_1 and y_2. Hence we have

$$y = y_1 + y_2 = A\left\{\sin\left[\frac{2\pi}{\lambda_1}(x - vt)\right] + \sin\left[\frac{2\pi}{\lambda_2}(x - vt)\right]\right\}$$

Since both waves have (we assume) the same velocity v, the combined disturbance moves like a structure of unchanging shape, just as a wave of a single wavelength is like a rigid sine curve moving along at speed v. The shape of the combination is most easily considered if we put $t = 0$; we then have

$$y = A\left[\sin\left(\frac{2\pi x}{\lambda_1}\right) + \sin\left(\frac{2\pi x}{\lambda_2}\right)\right]$$

Such a combination, for two wavelengths not very different from one another, is shown in Fig. 7–6. It looks precisely like a case of beats, as discussed in Chapter 2. Indeed, it *is* a beat phenomenon,

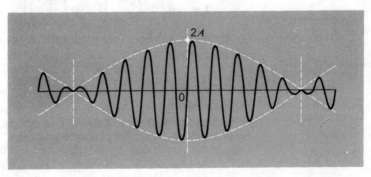

Fig. 7–6 Super-position of two travel-ing waves of slightly different wavelength.

although the modulation of amplitude is here a function of position instead of time. In discussing such superposed waves (and in other connections, too) it is extremely convenient to introduce the reciprocal of the wavelength. This quantity $k \, (= 1/\lambda)$ is called the *wave number*; it is the number of complete wavelengths per unit distance (and need not, of course, be an integer).[1]

In terms of wave numbers, the equation for the superposed wave form can be written as follows:

$$y = A[\sin 2\pi k_1 x + \sin 2\pi k_2 x]$$

or

$$y = 2A \cos [\pi(k_1 - k_2)x] \sin \left(2\pi \frac{k_1 + k_2}{2} x \right) \qquad (7\text{–}15)$$

The distance from peak to peak of the modulating factor is defined by the change of x corresponding to an increase of π in the quantity $\pi(k_1 - k_2)x$. Denoting this distance by D, we have

$$D = \frac{1}{k_1 - k_2} = \frac{\lambda_1 \lambda_2}{\lambda_2 - \lambda_1}$$

If the wavelengths are almost equal, we can write them as $\lambda, \lambda + \Delta\lambda$, and thus we have (approximately)

$$D \approx \frac{\lambda^2}{\Delta\lambda}$$

This means that a number of wavelengths given approximately by $\lambda/\Delta\lambda$ is contained between successive zeros of the modulation envelope.

The production of such superposed traveling waves on a string can be brought about by imposing two different frequencies and amplitudes of vibration simultaneously at one end of the string. This is expressed mathematically by considering the situation at $x = 0$ for the displacements defined by equations (7–14). We then have

$$y_0(t) = -A\left[\sin\left(\frac{2\pi v t}{\lambda_1}\right) + \sin\left(\frac{2\pi v t}{\lambda_2}\right)\right]$$

The ratio $2\pi v/\lambda$ defines the angular frequency ω of each vibration, and so we have

$$y_0(t) = -A[\sin \omega_1 t + \sin \omega_2 t]$$

---

[1] *Warning!* Because the combination $2\pi/\lambda$ occurs extremely frequently in the mathematical description of waves, it has become a common practice in theoretical physics to use the phrase "wave number" and the symbol k to designate this combination, which is equal to $2\pi k$ in our present notation.

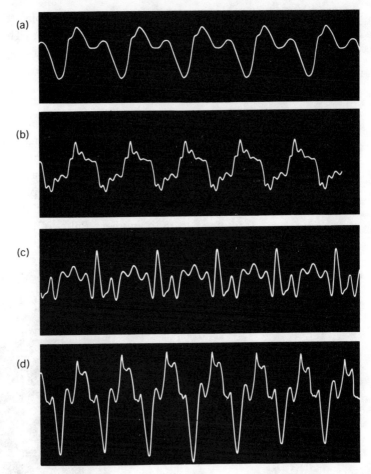

Fig. 7-7 Waveforms of (a) Flute. (b) Clarinet. (c) Oboe. (d) Saxophone. (From D. C. Miller, Sound Waves and Their Uses, Macmillan, New York, 1938.)

This then is an explicit case of beats in time, and we see here a particular example of the way in which a time-dependent disturbance at the source generates a space-dependent disturbance in the medium.

This superposition of waves is particularly beautifully illustrated by sound waves. In the transmission of sound from a source to a receiver we have a dual application of the principle just quoted. At the source there is some variation of displacement with time, as a result of which a train of sound waves is set up and travels away from the source. At some later time these waves, or some portion of them, fall upon a detector, producing in it a time-dependent displacement which, ideally, has exactly the same form as that which occurred at the source. Figure 7-7 shows some choice examples, and illustrates the way in which the

215   Superposition

harmonics of a given instrument combine to generate a pattern that repeats itself over and over again. The patterns represent the response of the receiver, but we can imagine at any instant a disturbance of the air, periodic in distance, to which the received signal corresponds.

WAVE PULSES

You may think of a wave as something that involves a whole succession of crests and troughs, but this is not at all necessary. Indeed, innumerable situations occur in which a single, isolated pulse of disturbance travels from one place to another through a medium—e.g., a single word of greeting or command shouted from one person to another. Pulses of this sort can be set up by taking a stretched spring (or elastic string) and producing in it a

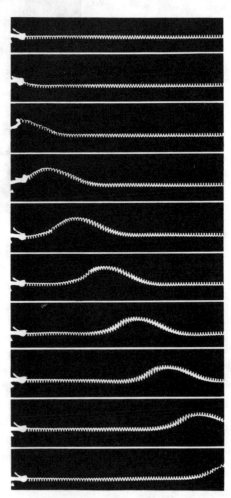

Fig. 7–8 Generation and motion of a pulse along a spring, shown by a series of pictures taken with a movie camera. (From Physical Science Study Committee, Physics, Heath, Boston, 1965.)

216 Progressive waves

local deformation—e.g., by twitching one end and then holding it still. Figure 7–8 shows the subsequent behavior of such a pulse. It travels along at a constant speed, so that at any instant only a limited region of the spring is disturbed, and the regions before and behind are quiescent. The pulse will continue to travel in this way until it reaches the far end of the spring, at which point a reflection process of some sort will occur. As long as the pulse continues uninterrupted, however, it appears to preserve the same shape, as Fig. 7–8 shows. How can we relate the behavior of such pulses to what we have already learned of sinusoidal waves? The answer is provided by Fourier analysis, and in the following discussion we shall see how this connection can be made. It is a very rewarding study, because it frees one to consider the transmission of any signal whatsoever.

Let us imagine first that we have an immensely long rope and that we oscillate one end up and down in simple harmonic motion with a period of 1 hr. To make things specific, let us suppose that the rope has a tension of 100 N and is of linear density 1 kg/m. Then the wave speed $\sqrt{T/\mu}$ is 10 m/sec, and the wavelength of our wave would be this speed v divided by the frequency ν ($= 1/3600 \text{ sec}^{-1}$) or, equivalently, the speed multiplied by the period (3600 sec), giving us $\lambda = 36,000$ m or about 22 miles! Let us imagine that our rope is several times longer than this—say 100 miles altogether. This particular arrangement is physically absurd, of course, but the consideration of it will help us to develop the essential ideas.

Suppose now that we oscillate the end of the rope with a combination of harmonics of the basic frequency. The second harmonic would generate sine waves of wavelength 18,000 m, the twenty-second harmonic would generate waves of wavelength about 1 mile, and the 36,000th harmonic would generate waves of wavelength 1 m. We cite these as specific examples, but the main point is that we can envisage the possibility of superposing thousands upon thousands of different sinusoidal vibrations at the driving end of the rope, all of them integral multiples of the same basic (and extremely low) frequency, and all giving rise to waves traveling along the rope at the same speed. And in consequence of this we would have, moving along the rope, a repeating pattern of disturbance, basically similar to those shown in Figs. 7–6 and 7–7, but in which the repetition distance was enormously long—and equal, in fact, to the wavelength associated with the basic frequency of 1 hr^{-1}.

But now let us introduce the remarkable possibilities implied in Fourier's theorem. Its claim is that, as we saw in Chapter 6 [Eq. (6–30)] *any* time-dependent pattern of displacement that repeats itself periodically (with a periodicity of $2\pi/\omega_1$) can be expressed as a linear combination of the infinite set of harmonics represented by ω_1 and all its integral multiples:

$$y(t) = \sum_{n=1}^{\infty} C_n \cos(n\omega_1 t - \delta_n) \qquad (7\text{–}16)$$

And the converse of this is that we can synthesize any repetitive pattern we like by means of the complete spectrum of harmonics of the basic frequency $\omega_1/2\pi$.

In particular, now, we can imagine a disturbance which is *zero* over most of the repetition period; some examples are shown in Fig. 7–9. According to Fourier's theorem, each of these, and any other such repetitive function of time, can be constructed from sinusoidal vibrations which, individually, are ever-continuing functions of time. The absence of any displacement over most of the repetition period $2\pi/\omega_1$ is brought about by just the right combination of harmonics, resulting in complete cancellation in this region, but nevertheless building up to give the particular nonzero disturbance over part of the period, as desired.

It will be noted that Eq. (7–16) [which is identical with Eq. (6–30)] implies that both sine and cosine functions of $n\omega_1 t$ are needed for the representation of an arbitrary periodic function, for we have

$$C_n \cos(n\omega_1 t - \delta_n) = A_n \sin n\omega_1 t + B_n \cos n\omega_1 t$$

Fig. 7–9 Examples of periodically repeated distur-bances, zero over most of the repetition period.

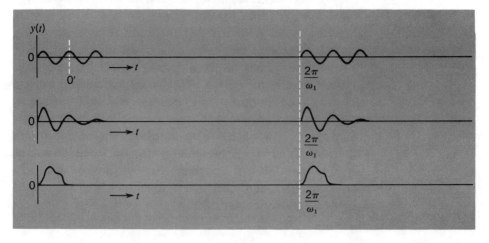

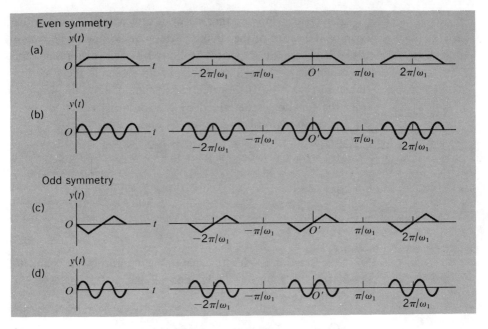

Even symmetry

(a)

(b)

Odd symmetry

(c)

(d)

Fig. 7-10 Shifting origin to achieve symmetry in various types of pulse.

Certain forms of $y(t)$ will, however, be describable in terms of sine functions or cosine functions alone. Specifically, if $y(t)$ is an *even* function of t, so that $f(-t) = +f(t)$ for any t, then the Fourier analysis requires cosine functions only; whereas if it is an *odd* function, so that $f(-t) = -f(t)$, then sine functions only will suffice. This kind of simplification will always be possible if the function $y(t)$ has odd or even symmetry with respect to its midpoint in time. One may, however, have to shift the origin of t to exploit this symmetry. Thus, for example, in Fig. 7-9(a) the function $y(t)$, consisting of $2\frac{1}{2}$ cycles of a sine wave followed by zero disturbance, is neither odd nor even with respect to the time origin shown. On the other hand, if the origin is shifted to the point O', corresponding to the central crest of the sine wave train, the function then is an even function with respect to O'. Similarly, any *whole* number of cycles of a sine wave, repeated at regular intervals, could be represented as an *odd* function through the appropriate shift of origin. In such cases a single repetition period is most conveniently measured between $t = -\pi/\omega_1$ and $t = +\pi/\omega_1$, rather than between 0 and $2\pi/\omega_1$. Figure 7-10 illustrates the application of this procedure to typical even or odd pulses.

219   Wave pulses

Example.[1] Suppose that we want to generate a wave in the form of 100 cycles of the 1000th harmonic—occupying one tenth of the basic repetition period—followed by zero disturbance for the other 90% of the time. This would resemble the situation shown in Fig. 7–10(d). As referred to the midpoint of the wave train the function is described by the following equations over the repetition period between $-\pi/\omega_1$ and $+\pi/\omega_1$:

$$y(t) = A_0 \sin N\omega_1 t \qquad 0 \le |t| \le \frac{100\pi}{N\omega_1}$$

$$y(t) = 0 \qquad \frac{100\pi}{N\omega_1} < |t| \le \frac{\pi}{\omega_1}$$

(7–17)

where

$$N = 1000$$

Since the function is odd, it is analyzable in terms of the complete set of functions $\sin n\omega_1 t$ only [i.e., all the phase angles δ_n in Eq. (7–16) are equal to $\pi/2$]:

$$y(t) = \sum_{n=1}^{\infty} C_n \sin n\omega_1 t \qquad (7\text{–}18)$$

and the coefficients C_n are obtained through the exploitation of the orthogonality of the sine functions with respect to integration over a complete period $2\pi/\omega_1$:

$$\int_{-\pi/\omega_1}^{\pi/\omega_1} \sin n_1\omega_1 t \sin n_2\omega_1 t \, dt = \begin{cases} 0 & \text{if } n_1 \ne n_2 \\ \dfrac{\pi}{\omega_1} & \text{if } n_1 = n_2 \end{cases}$$

Hence we have, after multiplying Eq. (7–18) by $\sin n\omega_1 t$ and integrating, the result

$$C_n = \frac{\omega_1}{\pi} \int_{-\pi/\omega_1}^{\pi/\omega_1} y(t) \sin n\omega_1 t \, dt$$

In this we substitute for $y(t)$ as given by equations (7–17), which therefore gives us

$$C_n = \frac{\omega_1 A_0}{\pi} \int_{-100\pi/N\omega_1}^{100\pi/N\omega_1} \sin N\omega_1 t \sin n\omega_1 t \, dt$$

(Note that the limits of integration are now $\pm 100\pi/N\omega_1$, because outside these limits the integrand is zero.) Let us evaluate this integral by using the relation

$$\sin N\omega_1 t \sin n\omega_1 t = \tfrac{1}{2}[\cos(N - n)\omega_1 t - \cos(N + n)\omega_1 t]$$

[1]This may be skipped without any loss of continuity.

220 Progressive waves

Therefore,

$$\int \sin N\omega_1 t \sin n\omega_1 t \, dt = \tfrac{1}{2}\left[\frac{\sin(N-n)\omega_1 t}{(N-n)\omega_1} - \frac{\sin(N+n)\omega_1 t}{(N+n)\omega_1}\right]$$

Inserting the limits on t, we see that $\omega_1 t$ takes on the values $\pm 100\pi/N$. Hence we have

$$C_n = \frac{\omega_1 A_0}{\pi}\left\{\frac{\sin\left[\dfrac{100\pi(N-n)}{N}\right]}{(N-n)\omega_1} - \frac{\sin\left[\dfrac{100\pi(N+n)}{N}\right]}{(N+n)\omega_1}\right\}$$

Here we shall introduce an approximation. We note that the first term inside the braces develops a small denominator for $n \approx N$, whereas the denominator of the second term is always large. The maximum possible value of the numerator in each is unity. Thus it is possible for most purposes to ignore the second term, which allows us to write a simplified approximate expression for the amplitudes C_n:

$$C_n \approx \frac{A_0}{\pi}\left\{\frac{\sin\left[\dfrac{100\pi(N-n)}{N}\right]}{N-n}\right\}$$

or

$$C_n \approx \frac{100A_0}{N}\left(\frac{\sin\theta_n}{\theta_n}\right) \quad \text{where } \theta_n = \frac{100\pi(N-n)}{N}$$

These values of C_n are sizable only in the neighborhood of $n = N$. The function $(\sin\theta_n)/\theta_n$ is unity at $\theta_n = 0$ and falls to zero at $\theta_n = \pm\pi$ (beyond which it oscillates through negative and then positive values with steadily decreasing amplitude).[1] If $N = 1000$, as we have assumed, then $\theta_n = \pm\pi$ at $n = N \pm 10$. And what this means is that the spectrum of our group of 100 cycles of $N = 1000$ is, primarily, a cluster of contributions as shown in Fig. 7–11, with $n = 1000$ itself providing the biggest single amplitude.

If we allowed our chosen vibration to continue for a larger number of cycles, its spectrum in terms of the pure harmonics, indefinitely maintained, would narrow down until, in the limit of infinitely many cycles, we would, of course, be left with the single pure harmonic $N = 1000$ all by itself. On the other hand, a pulse made up of only a few cycles of a given harmonic fre-

[1] The appearance of *negative* values of C_n can, as in our discussion of the forced oscillator, be described by a phase change of π. One could, therefore, describe these contributions in terms of positive values of C_n associated with phases of δ_n equal to $3\pi/2$ (or $-\pi/2$) instead of $\pi/2$.

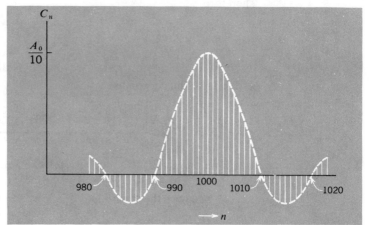

Fig. 7–11 Frequency spectrum (amplitude plotted against frequency, as obtained by Fourier analysis) for a signal consisting of 100 cycles of a pure sine wave repeated at time intervals of 1000 cycles.

quency would require the use of an exceedingly broad spectrum (i.e., many harmonics with comparable amplitudes) in its Fourier synthesis.[1] This essentially inverse connection between the duration of a pulse and the width of its frequency spectrum is a very fundamental one. It is precisely this kind of result that we drew attention to in Chapter 1, giving in effect a warning that perfectly pure sinusoidal disturbances do not really exist. But, of course, a sinusoidal vibration that continues for, let us say, a million cycles is very close indeed to having a frequency spectrum consisting of a single sharp line.

Let us return now to the more qualitative aspects of a repeated vibration with long intervals of quiescence between-times. We regard this vibration, caused at a given place, as being analyzed into its complete spectrum of Fourier components. Provided, now, that the wave speed associated with each component frequency is precisely the same, these intermittent but periodically repeated vibrations will give rise to isolated pulses, equally spaced, traveling through the medium. With our very long rope, for example, one could imagine the possibility of generating wave pulses of the sort shown in Fig. 7–9(c), with an over-all length of a few meters, and separated by the basic repetition distance of 36 km. To all intents and purposes these would be isolated, individual disturbances. It does not take much imagination, in fact, to see that the principles of Fourier analysis can be pushed to a limit in which the repetition period is infinitely long, and so, therefore, is the repetition distance of a waveform in the traveling wave. We can thus envisage the description of

[1]You should satisfy yourself that the preceding analysis implies this property.

222   Progressive waves

one single, nonrepeated pattern of displacement as a function of time or position, in terms of a complete (continuous) spectrum of sinusoidal disturbances with periods or wavelengths extending up to infinitely large values.

It is in the above terms, then, and subject to the condition that the speed of pure sinusoidal waves is independent of their frequency or wavelength, that we can envisage the propagation, without any change of shape, of arbitrary isolated pulses through a medium. Let us now consider some features of the motion of such pulses.

MOTION OF WAVE PULSES OF CONSTANT SHAPE

Given a pulse that satisfies the conditions discussed above, we can proceed to discuss its behavior in quite general terms. Suppose that a pulse is moving from left to right, and that at a time we shall call $t = 0$ it is described by a certain equation:

$$y_{(t=0)} = f(x)$$

If the pulse as a whole is traveling at a velocity v, then at a later time t the displacement that originally existed at some particular value of x (say x_1) is found to be now at x_2, where

$$x_2 = x_1 + vt$$

The equation of the pulse at this new value of t can be obtained by recognizing that a picture of the pulse at time t looks just the same as a picture at $t = 0$ except for a shift of the origin of x by the distance vt (see Fig. 7–12). We can express this mathematically by saying that the transverse displacement, for any values of x and t, is given by

$$y(x, t) = f(x - vt) \tag{7–19}$$

The choice of this analytic form can be verified, just as for the particular case of a pure sine wave, by considering the condition for a particular value of y to be found at $(x + \Delta x, t + \Delta t)$ after being previously observed at (x, t). In similar fashion a pulse traveling from right to left is described by

$$y(x, t) = g(x + vt) \tag{7–20}$$

Fig. 7–12 Movement of an arbitrary traveling pulse.

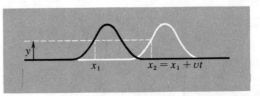

x_1 $x_2 = x_1 + vt$

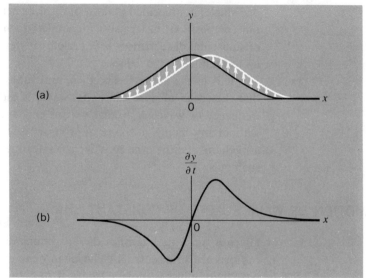

The exact form of the functions f and g is immaterial. All that matters is that y should be expressible as a function of $x \pm vt$. Thus, for example, we could define a certain shape of pulse, moving from left to right, by the equation

$$y(x, t) = \frac{b^3}{b^2 + (x - vt)^2} \qquad (7\text{–}21)$$

Sketches of this pulse for $t = 0$ and for a slightly later time are shown in Fig. 7–13(a). The peak of the pulse would be of height b, and this peak would pass through the point $x = 0$ at $t = 0$. The pulse would fall to half-maximum height at the points $x = vt \pm b$ and would be down to less than 10% of its peak height for $|x - vt| > 3b$. And one could write down any number of other possible pulse shapes, using powers, exponentials, trigonometric functions, etc. But all such pulses travel in the same way, preserving their shape and moving at the same speed v, if they are correctly described by one or the other of Eqs. (7–19) and (7–20).

It is very important for an understanding of waves to appreciate how the motion of a wave profile along its direction of propagation (x) can be the consequence of particle displacements that are purely along a transverse direction (y). Thus, for example, the pulse of Fig. 7–13(a) moves to the right because, at any instant, the transverse displacement of every point to the left of the peak is decreasing and the displacement of every point to the right of the peak is increasing. It is an automatic consequence of these motions that the peak displacement occurs at larger and larger values of x as time goes on.

Let us calculate the distribution of transverse velocities for the pulse described by Eq. (7–21). The transverse velocity of any particle of the medium (spring, string, or whatever) is the rate of change of y with t at some given value of x, i.e.,

$$v_y = \frac{\partial y}{\partial t}$$

where we use the partial derivative notation, recognizing that y is a function of both x and t and that we are holding x fixed. Thus, from Eq. (7–21) we have

$$v_y = \frac{-b^3}{[b^2 - (x - vt)^2]^2} \frac{\partial}{\partial t} [b^2 + (x - vt)^2]$$

i.e.,

$$v_y(x, t) = \frac{2b^3(x - vt)v}{[b^2 + (x - vt)^2]^2} \tag{7-22}$$

This defines the transverse velocity at any point at any time. Suppose now that we want the distribution of transverse velocities at $t = 0$, when the peak of the pulse is passing through the point $x = 0$. Putting $t = 0$ in Eq. (7–22) we have

$$v_y(x, 0) = \frac{2b^3vx}{(b^2 + x^2)^2}$$

The graph of this velocity distribution is shown in Fig. 7–13(b), and it is easy to see how these velocities, operating for a short time Δt, give rise to small vector displacements that shift the pulse as a whole in the way indicated in Fig. 7–13(a). It must be recognized, of course, that the velocity distribution itself moves with the pulse, so that the condition $v_y = 0$ is always satisfied at the peak of the pulse. The form of Eq. (7–22) embodies this condition, because it shows that v_y, like y itself, is a function of the combined variable $x - vt$.

You may have recognized already that there is an intimate connection between the transverse velocity and the slope of the pulse profile. For suppose (see Fig. 7–14) that an instantaneous picture of a pulse shows a small portion of it to be along the straight line AB. The slope can be measured as $A'B/AA'$. But in some short interval of time Δt the line AB would move to $A'B'$; this time is given by

$$\Delta t = \frac{AA'}{v}$$

Fig. 7-14 Relation between transverse displacement of a medium and longitudinal displacement of a traveling pulse.

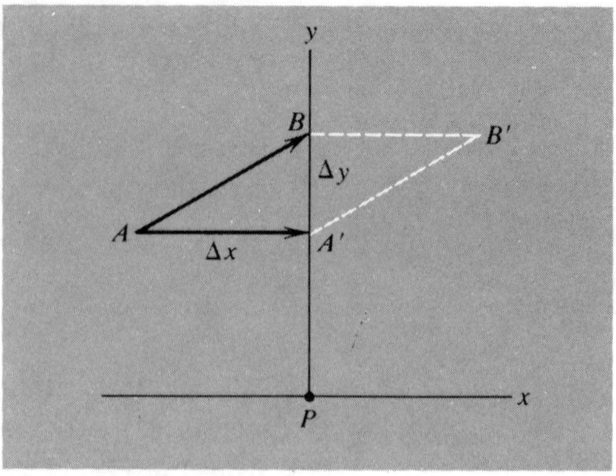

where v is the velocity with which the pulse travels. If, however, we confined our observations to the particular value of x indicated by the vertical line, we should see the transverse displacement change from PB to PA' as the pulse passed by. The amount of this displacement is thus just the negative of the distance $A'B$, and the associated transverse velocity is

$$-A'B/\Delta t = -v(A'B/AA').$$

Let us express this in the language of partial derivatives. The slope $A'B/AA'$ is the value of $\Delta y/\Delta x$ at some fixed value of t, and from the above discussion we can see that (in the limit) the following relation holds:

$$v_y = -v\frac{\partial y}{\partial x}$$

Since v_y is the value of $\Delta y/\Delta t$ at some fixed value of x, we can alternatively write this as

$$v_y = \frac{\partial y}{\partial t} = -v\frac{\partial y}{\partial x}$$

Thus the transverse velocity at any point is directly proportional to the slope of the pulse profile at that point.

We can complete this analysis by recalling that v itself is defined as the limiting value of $\Delta x/\Delta t$ for some fixed value of y, i.e.,

$$v = \frac{\partial x}{\partial t}$$

Putting all these together gives the following result:

$$v_y = \frac{\partial y}{\partial t} = -\frac{\partial y}{\partial x}\frac{\partial x}{\partial t} \tag{7-23}$$

Equation (7–23) is deceptively like the chain rule for ordinary differentiation—but notice the minus sign. What we have here is a special case of a more general kind of situation, in which some quantity y is a function of both position and time. It may vary from place to place at a given instant, and it may vary with time at a given place. Two successive observations of y, separated by a time Δt, and at positions separated by Δx, then differ by an amount Δy which can be expressed as follows:

$$\Delta y = \frac{\partial y}{\partial t}\Delta t + \frac{\partial y}{\partial x}\Delta x$$

The over-all rate of change of y is thus given by

$$\frac{dy}{dt} = \frac{\partial y}{\partial t} + v\frac{\partial y}{\partial x} \tag{7-24}$$

where v is the velocity $\Delta x/\Delta t$. The operator $\partial/\partial t + v\partial/\partial x$ is often called the *convective derivative*. It defines the way of obtaining the time rate of change of y if one's point of observation is being moved along at some defined velocity—as for example, through the bodily movement of a fluid. And if, in Eq. (7–24), one inserts the condition $dy/dt = 0$, this corresponds to fixing attention on a particular value of y, just as we have indeed done in defining the motion of a point of given displacement in an arbitrary pulse profile. But this condition—$dy/dt = 0$—then converts Eq. (7–24) into the special statement expressed in Eq. (7–23).

It is easy to see that our general equations, Eqs. (7–19) and (7–20), both satisfy the same basic differential equation of wave motion. [We have, of course, really assured ourselves of this in advance, by first recognizing that any such traveling pulse is a superposition of sinusoidal waves that all obey Eq. (7–9).] We have the two equations

$$y(x, t) = \begin{cases} f(x - vt) \\ g(x + vt) \end{cases}$$

For the first of them, we have

$$\frac{\partial y}{\partial x} = \frac{df}{d(x - vt)} \frac{\partial(x - vt)}{\partial x} = f'$$

where f' is the derivative of f with respect to the whole argument $(x - vt)$. Differentiating again,

$$\frac{\partial^2 y}{\partial x^2} = f''$$

where f'' is the second derivative of f with respect to $(x - vt)$. Differentiating now with respect to t,

$$\frac{\partial y}{\partial t} = f' \frac{\partial(x - vt)}{\partial t} = -vf'$$

And, after a second differentiation,

$$\frac{\partial^2 y}{\partial t^2} = (-v)^2 f'' = v^2 f''$$

Comparing these two second derivatives, we see that

$$\frac{\partial^2 y}{\partial t^2} = v^2 \frac{\partial^2 y}{\partial x^2}$$

which thus reproduces Eq. (7–9). And if we go through the same procedure with the function $g(x + vt)$, which describes an arbitrary disturbance traveling in the negative x direction, the only difference is that a factor $+v$, instead of $-v$, appears as a result of each differentiation with respect to t. Thus after two differentiations, the functions f and g are seen to obey the same equation.

SUPERPOSITION OF WAVE PULSES

In the last section we limited ourselves to the consideration of individual pulses. But one of the most important and interesting features of the behavior of such pulses is that two of them, traveling in opposite directions, can pass right through each other and emerge from the encounter with their separate identities.

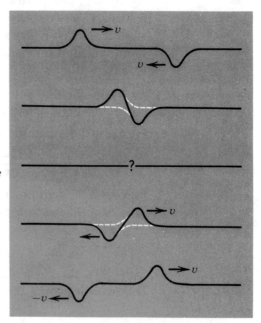

Fig. 7–15 Successive superposition of two pulses that are reversed right to left and top to bottom with respect to one another and that travel in opposite directions.

This is superposition at work once again, in a very remarkable form. Figure 7–15 shows what is perhaps the most surprising type of such superposition. Two symmetrical pulses are traveling in opposite directions; they are exactly alike, except that one is positive and the other is negative. As they pass through each other, there comes a moment at which the whole spring or string is straight; it is as if the pulses had annihilated each other, and so, in a sense, they have. But your intuitions will tell you that each pulse was carrying a positive amount of energy, which cannot simply be washed out. And, indeed, the pulses do reappear.[1] But what is it that preserves the memory of them through the stage of zero displacement, so that they are recovered intact in their original form? It is the velocity of the different parts of the system. The string at the instant of zero transverse deformation has a distribution of transverse velocities characteristic of the two superposed pulses—and the velocity distribution of a symmetrical positive pulse traveling to the right is exactly the same as that of a similar negative pulse traveling to the left. This is implied by Eq. (7–23)—since reversing the signs of both $\partial y/\partial x$ and $\partial x/\partial t$ leaves v_y unchanged—but is also immediately apparent if one makes a sketch of the two pulses as they appear at two successive instants. Thus the transverse displacements cancel, but the transverse velocities add, and for this one instant the whole energy

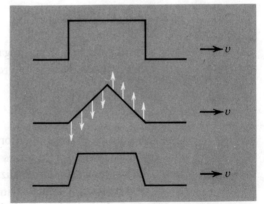

Fig. 7–16 Geometri-
cal idealizations of
simple types of pulse.

[1]Leonardo da Vinci, one of the keenest observers of all time, studied waves extensively and recognized the results of such superposition, but did not discern the mechanism. Thus he wrote: "All the impressions caused by things striking upon the water can penetrate one another without being destroyed. One wave never penetrates another; but they only recoil from the spot where they strike." See *The Notebooks of Leonardo da Vinci*, translated by Edward McCurdy, Braziller, New York, 1956.

229   Superposition of wave pulses

of the system resides in the kinetic energy associated with these velocities. But let us concentrate for the moment on the purely kinematic aspects of the problem.

It may for some purposes be convenient to assume simple geometric shapes for pulse profiles—such as the rectangle, triangle, and trapezoid shown in Fig. 7–16. With a triangular pulse, for example, the transverse velocity is the same for all points along each side of the pulse, and the consequences of superposing such pulses are easily analyzed. It should be realized, however, that such shapes are unphysical. Thus the passage of a rectangular pulse would require the transverse velocity to be infinitely great as the vertical sides of the pulse passed by. And any pulse profile with sharp corners (such as the trapezoid) implies discontinuous changes in transverse velocity, which in turn means infinite accelerations requiring infinite forces. Any real pulse, therefore, has rounded corners and sloping sides, however exotic its shape may be otherwise.

DISPERSION; PHASE AND GROUP VELOCITIES

We have given the equation of a progressive sine wave in the form [Eq. 7–7)]

$$y(x, t) = A \sin\left[\frac{2\pi}{\lambda}(x - vt)\right]$$

For a stretched string, regarded as having a continuous distribution of mass, we had the relation [Eq. (7–5)]

$$v = \left(\frac{T}{\mu}\right)^{1/2}$$

According to these equations, a given string, under a given tension, will carry sinusoidal waves of all wavelengths at the same speed v. This is, however, an idealization which will certainly fail, to some degree, for any actual string. We pointed to this limitation most particularly in Chapter 5, in our discussion of the normal modes of a line of connected masses. What emerged there was that for a lumpy string of length L, fixed at its ends, the wavelength λ_n that could be associated with a given normal mode, n, was $2L/n$—just as for a continuous string—but that the mode frequency ν_n was not simply proportional to n. Instead, the mode frequency was found to be given by

$$\nu_n = 2\nu_0 \sin\left[\frac{n\pi}{2(N + 1)}\right]$$

so that the value of $2\nu_0$ defined an upper limit to the possible frequency of any line made up of a finite number (N) of masses [see Eq. (5–25) p. 141]. For $n \ll N$, this reduced to the same result as for a continuous string, with ν_n proportional to n. But with increasing n, the values of ν_n would rise less and less rapidly than this proportionality would require.

In general, therefore, we must expect that, for waves on a string, pure sinusoidal waves of high frequency and short wavelength tend to travel with smaller speeds than the longer waves. This is one example of what is called *dispersion*, a variation of wave speed with wavelength.

The phenomenon of dispersion is to be found in many different kinds of media, with different underlying physical mechanisms. And what we want to stress is not the very special analysis that led us to the dispersive property of a string of beads, but the fact of dispersion itself. The word suggests a separation of what was at first in one place, and that is exactly what it entails. We see it happening when white light passes through a prism and is spread out into its different colors. The velocity for waves of red light in glass is greater than that for waves of blue light, and the refraction of light upon entering the prism is given by Snell's law:

$$\frac{\sin i}{\sin r} = n = \frac{c}{v}$$

so that the angle of refraction varies with the color according to the variation of velocity. In a one-dimensional problem the dispersion would mean that two long but limited trains of waves, of different wavelengths, would get further apart as time went on, if initially they overlapped. Also each individual wave train, being itself an admixture of pure sine waves of slightly different velocities, would become distorted and more spread out with the passage of time. Only a pure sine wave of effectively infinite extent, with a unique wavelength and frequency, would move with a uniquely defined velocity in a dispersive medium. (Of course, the dispersion *may* be negligible in particular circumstances—and for the special case of light waves in vacuum it appears to be strictly zero.)

To discuss the consequences of dispersion more concretely, we shall consider what happens if we have two sinusoidal waves of slightly different wavelengths traveling in the same direction

231   Dispersion; phase and group velocities

(but perhaps at different speeds) along a string. Suppose for simplicity that they have equal amplitudes, and that they are described by the following equations:

$$y_1 = A \sin 2\pi(k_1 x - \nu_1 t)$$
$$y_2 = A \sin 2\pi(k_2 x - \nu_2 t)$$

(7-25)

These are very much like the equations (7-14) that we wrote down in order to calculate the waveform of two waves having the same velocity. For convenience in handling the equations, however, we are using the wave number k instead of $1/\lambda$, and we are explicitly inserting the frequency ν in place of the ratio v/λ. In general, now, we are supposing that these two waves have *different* characteristic speeds:

$$v_1 = \frac{\nu_1}{k_1} = \nu_1 \lambda_1 \qquad v_2 = \frac{\nu_2}{k_2} = \nu_2 \lambda_2$$

The superposition of these two waves gives us a combined disturbance as follows:

$$y = A[\sin 2\pi(k_1 x - \nu_1 t) + \sin 2\pi(k_2 x - \nu_2 t)]$$

Using the same trigonometric relations as we employed before, this becomes

$$y = 2A \cos \pi[(k_1 - k_2)x - (\nu_1 - \nu_2)t]$$
$$\times \sin 2\pi \left[\frac{k_1 + k_2}{2} x - \frac{\nu_1 + \nu_2}{2} t \right]$$

At $t = 0$ this looks just like the superposed waves of Fig. 7-6. But now let us consider what happens with the passage of time. The above expression for y can be interpreted as a rapidly alternating wave of short wavelength, modulated in amplitude by an envelope of long wavelength. Both of these wavelike disturbances move. *But they may have different speeds.* A place of maximum possible amplitude necessarily moves at the speed of the envelope.

If the two combining waves are of almost the same wavelength, we can simplify our description of the combined disturbance by putting

$$k_1 - k_2 = \Delta k \qquad \nu_1 - \nu_2 = \Delta \nu$$
$$\frac{k_1 + k_2}{2} = k \qquad \frac{\nu_1 + \nu_2}{2} = \nu$$

Then we get

$$y = 2A \cos \pi(x \Delta k - t \Delta \nu) \sin 2\pi(kx - \nu t)$$

(7-26)

In this expression we can then identify two characteristic velocities. One of these is the speed with which a crest belonging to the average wave number k moves along. This is called the *phase velocity*, v_p:

$$v_p = \frac{\nu}{k} = \nu\lambda \tag{7-27}$$

The other is the velocity with which the modulating envelope moves. Because this envelope encloses a group of the short waves, the velocity in question is called the *group velocity*, v_g:

$$v_g = \frac{\Delta\nu}{\Delta k} \longrightarrow \frac{d\nu}{dk} \tag{7-28}$$

The phase velocity is the only kind of velocity that we have associated with a wave up till now. It is given this name because it represents the velocity that we can associate with a fixed value of the phase in the basic shortwave disturbance—e.g., representing the advance of x with t for a point of zero displacement.

The group velocity is of great physical importance, because every wave train has a finite extent, and except in those rare cases where we follow the motion of an individual wave crest, what we observe is the motion of a wave group. Also, it turns out that the transport of energy in a wave disturbance takes place at the group velocity. To treat such questions effectively one needs to use, not just two sine waves, but a whole spectrum, sufficient to define a single isolated pulse or wave group, in the manner we discussed earlier. When this is done, the value of the group velocity is still found to be given by Eq. (7–28).

The existence of dispersion does, of course, carry important implications for this matter of analyzing an arbitrary pulse into pure sinusoids. If these sinusoids have different characteristic speeds, the shape of the disturbance must change as time goes on. In particular, a pulse that is highly localized initially will suffer the fate of becoming more and more spread out as it moves along.

A striking example of the difference between phase and group velocities is provided by waves in deep water—so-called "gravity waves." These are strongly dispersive; the wave speed for a well-defined wavelength—what we must now call the phase velocity—is proportional to the square root of the wavelength. Thus we can put

$$v_p = C\lambda^{1/2} = Ck^{-1/2}$$

where C is a constant. But $v_p = \nu/k$, by Eq. (7–27). Hence we have

$$\nu = Ck^{1/2}$$

233   Dispersion; phase and group velocities

Therefore,

$$\frac{dv}{dk} = \tfrac{1}{2}Ck^{-1/2}$$

But dv/dk is the group velocity, and thus we have

$$v_g = \tfrac{1}{2}v_p$$

so that the component wave crests will be seen to run rapidly through the group, first growing in amplitude and then apparently disappearing again. You may have noticed this curious effect on the surface of the sea or some other body of deep water.

Sound waves in gases, like the other elastic vibrations we have considered, are nondispersive—at least, to the extent that our theoretical description is correct. This is a fortunate circumstance. Imagine the chaos and aural anguish that would result if sounds of different frequencies traveled at different speeds through the air. Listening to an orchestra could be a veritable nightmare. Of course, it would have its compensations—we could, for example, analyze sounds with a prism of gas, just as we can analyze light with a prism of glass. But as human beings we can be content that this possibility does not offer itself.

THE PHENOMENON OF CUT–OFF[1]

Closely linked to the property of dispersion is the very remarkable effect known as cut-off. This term describes the inability of a dispersive medium to transmit waves above (or possibly below) a certain critical frequency. The effect is implicit in the analysis of the normal modes of a line of N separated masses, for which we found [see Eq. (5–24), p. 141]

$$\omega_n = 2\omega_0 \sin\left(\frac{n\pi l}{2L}\right)$$

where $L = (N + 1)l$. We can imagine that the length L of the line is increased indefinitely, without changing the separation l between adjacent masses. In this case the wave number k_n ($= n/2L$) becomes in effect a continuous variable, and we can write the relationship between frequency v and wave number k as follows:

$$v(k) = 2v_0 \sin(\pi k l) \tag{7–29}$$

Clearly Eq. (7–29) does not permit any value of $v(k)$

[1]This section can be omitted without loss of continuity.

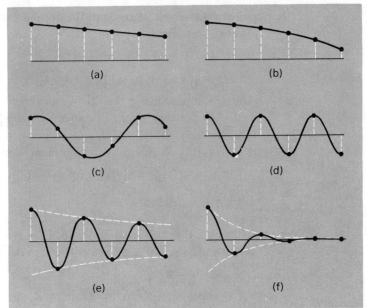

Fig. 7–17 Amplitude relationships for particles on a string, driven at left end. (a) Static equilibrium, $\nu = 0$. (b) $\nu \ll \nu_0$. (c) $\nu = \sqrt{2}\nu_0$. (d) Highest mode, $\nu = 2\nu_0$. (e) $\nu > 2\nu_0$ (f) $\nu \gg 2\nu_0$.

greater than $2\nu_0$. Thus we recognize (as already discussed on p. 142) the existence of a maximum normal mode frequency $\nu_m \,(= 2\nu_0 = \omega_0/\pi)$. This frequency ν_m corresponds to a wave number k_m such that

$$\pi k_m l = \pi/2$$

or to a wavelength λ_m equal to $2l$. But if we had such a line of masses, there would be nothing to prevent us from shaking one end at a frequency *greater* than ν_m. What, in fact, happens in this case?

To find out, we go back to the equation that relates the amplitudes of successive masses in the coupled system vibrating at some frequency ν (or ω). From Eq. (5–19) we have the following relationship between the amplitudes A_{p-1}, A_p, A_{p+1} for three successive particles (see p. 140):

$$\frac{A_{p-1} + A_{p+1}}{A_p} = \frac{-\nu^2 + 2\nu_0{}^2}{\nu_0{}^2} \qquad (7\text{--}30)$$

Let us consider the kind of picture that this equation gives us for various values of ν.

a. $\nu = 0$. In this case,

$$A_p = \tfrac{1}{2}(A_{p-1} + A_{p+1})$$

The amplitude varies linearly with distance along the line; it is a simple static equilibrium [Fig. 7–17(a)] with one end of the line pulled transversely aside from the normal resting position. The effective wavelength is infinite.

b. $\nu \ll \nu_0$. We now have

$$A_p > \tfrac{1}{2}(A_{p-1} + A_{p+1})$$

Any one amplitude is greater than the average of the two adjacent ones—but not very much. The effect is to produce a slight curvature, toward the axis, of a smooth curve joining the particles [Fig. 7–17(b)] which ensures a sinusoidal form.

c. $\nu = \sqrt{2}\,\nu_0$. This is a very special case. We now have

$$\frac{A_{p-1} + A_{p+1}}{A_p} = 0$$

Remember that this must be satisfied for every set of three consecutive masses, not just for a particular set. It requires

$$A_{p+1} = -A_{p-1}$$

but it appears to place no requirement on the ratio A_{p-1}/A_p. Thus the situation might be as indicated in Fig. 7–17(c). The wavelength associated with this frequency is clearly $4l$, where l is the interparticle distance. This conclusion is confirmed by Eq. (7–29), which for $k = 1/4l$ gives us

$$\nu = 2\nu_0 \sin\frac{\pi}{4} = \sqrt{2}\,\nu_0$$

d. $\nu = 2\nu_0$. This represents the maximum frequency ν_m for a normal mode. From Eq. (7–30) we have

$$A_p = -\tfrac{1}{2}(A_{p-1} + A_{p+1})$$

It requires an alternation of positive and negative displacements of the same size, as shown in Fig. 7–17(d) and as discussed near the end of Chapter 5. The wavelength is $2l$, again in conformity with Eq. (7–29).

e. $\nu > 2\nu_0$. Suppose that ν is greater than $2\nu_0$, but not very much greater. Then A_p is opposite in sign to the mean of A_{p-1} and A_{p+1}, and also

$$|A_p| < \tfrac{1}{2}|A_{p-1} + A_{p+1}|$$

This implies a slight curvature, *away* from the axis, of the smooth curves joining alternate particles. If it is the left hand of the line that is being shaken, we would be led to Fig. 7–17(e) as a reasonable representation of the displacements. The amplitudes alternate in sign, and fall off in magnitude in geometric proportion—i.e., exponentially. This is the phenomenon of cut-off.

Let us put $\nu = 2\nu_0 + \Delta\nu$, and let the ratios A_{p-1}/A_p, A_p/A_{p+1}, etc., be set equal to $-(1 + f)$, where f is some small fraction. From Eq. (7–30) we have

$$\frac{A_{p-1}}{A_p} + \frac{A_{p+1}}{A_p} = \frac{-\nu^2}{\nu_0{}^2} + 2$$

Therefore,

$$-(1 + f) - (1 + f)^{-1} = \frac{-(2\nu_0 + \Delta\nu)^2}{\nu_0{}^2} + 2$$

Therefore,

$$-1 - f - (1 - f + f^2 \cdots) = \frac{-[4\nu_0{}^2 + 4\nu_0\,\Delta\nu + (\Delta\nu)^2]}{\nu_0{}^2} + 2$$

i.e.,

$$-2 - f^2 + \cdots = -2 - \frac{4\,\Delta\nu}{\nu_0} - \left(\frac{\Delta\nu}{\nu_0}\right)^2$$

Hence, approximately,

$$f = 2\left(\frac{\Delta\nu}{\nu_0}\right)^{1/2}$$

The further we go above the critical frequency ν_m, the more drastic is the attenuation as we proceed along the line, as suggested by the comparison of Figs. 7–17(e) and (f).

f. $\nu \gg 2\nu_0$. This brings us to the situation of being far above the critical frequency of cut-off. It will now be very nearly correct to put

$$\frac{A_{p-1}}{A_p} = -\frac{\nu^2}{\nu_0{}^2}$$

Thus, for example, if $\nu = 2\nu_m = 4\nu_0$, it will be almost true to say that only the first particle in the line—the one being agitated by some external driving agency—will show any appreciable response; the rest of the line behaves almost as a rigid structure.

THE ENERGY IN A MECHANICAL WAVE

At any instant the particles of a medium carrying a wave are in various states of motion. Clearly the medium is endowed with energy that it does not have in its normal resting state. There are contributions from the potential energy of deformation as well as from the kinetic energy of the motion. We shall calculate the

237   The energy in a mechanical wave

total energy associated with one complete wavelength of a sinusoidal wave on a stretched string.

By way of approaching this problem, we shall consider first a small segment of the string—so short that it can be regarded as effectively straight—that lies between x and $x + dx$, as shown in Fig. 7–18. We shall make the usual assumptions that the displacements of the particles in the string are strictly transverse and that the magnitude of the tension T is not changed by the deformation of the string from its normal length and configuration.

The mass of the small segment is $\mu\, dx$, and its transverse velocity (u_y) is $\partial y / \partial t$. Hence, for this segment, we have

$$\text{kinetic energy} = \tfrac{1}{2}\mu\, dx \left(\frac{\partial y}{\partial t}\right)^2$$

and we can define a kinetic energy *per unit length*—what is called the kinetic-energy *density*—for such a one-dimensional medium:

$$\text{kinetic-energy density} \equiv \frac{dK}{dx} = \frac{1}{2}\mu\left(\frac{\partial y}{\partial t}\right)^2 \tag{7–31}$$

The potential energy can be calculated by finding the amount by which the string, when deformed, is longer than when it is straight. This extension, multiplied by the assumed constant tension T, is the work done in the deformation. Thus, for the segment, we have

$$\text{potential energy} = T(ds - dx)$$

where

$$ds = (dx^2 + dy^2)^{1/2}$$
$$= dx\left[1 + \left(\frac{\partial y}{\partial x}\right)^2\right]^{1/2}$$

If we assume that the transverse displacements are *small*, so that $\partial y / \partial x \ll 1$, we can approximate the above expression using the binomial expansion to two terms, thus getting

$$ds - dx \approx \frac{1}{2}\left(\frac{\partial y}{\partial x}\right)^2 dx$$

Therefore,

$$\text{potential energy} \approx \tfrac{1}{2}T\left(\frac{\partial y}{\partial x}\right)^2 dx$$

Hence we have

$$\text{potential-energy density} \equiv \frac{dU}{dx} \approx \frac{1}{2}T\left(\frac{\partial y}{\partial x}\right)^2 \tag{7–32}$$

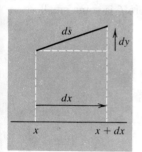

Fig. 7–18 *Displacement and extension of a short segment of string carrying a transverse elastic wave.*

It is worth noting that the kinetic-energy and potential-energy densities, as given by Eqs. (7–31) and (7–32), are *equal*. For, as we have seen, a traveling wave on the string is of the form

$$y(x, t) = f(x \pm vt) = f(z), \text{ say,}$$

where

$$v = \left(\frac{T}{\mu}\right)^{1/2}$$

Thus

$$\frac{\partial y}{\partial x} = f'(z)$$

$$\frac{\partial y}{\partial t} = \pm v f'(z)$$

Therefore,

$$\frac{dK}{dx} = \tfrac{1}{2}\mu v^2 [f'(z)]^2$$

$$\frac{dU}{dx} = \tfrac{1}{2}T[f'(z)]^2$$

which are equal since $T = \mu v^2$. Although this equality of the two energy densities cannot be assumed to hold good in all conceivable situations, it is in keeping with what we know about the equal division, on the average, of the total energy of simple mechanical systems subject to linear restoring forces.

Suppose now that we have, in particular, a sinusoidal wave described by the equation

$$y(x, t) = A \sin 2\pi v \left(t - \frac{x}{v}\right) \tag{7–33}$$

Then at any given value of x we have

$$u(x, t) = \frac{\partial y}{\partial t} = 2\pi v A \cos 2\pi v \left(t - \frac{x}{v}\right)$$

$$= u_0 \cos 2\pi v \left(t - \frac{x}{v}\right)$$

where $u_0 \ (= 2\pi v A)$ is the maximum speed of the transverse motion. Let us consider this distribution of transverse velocities at the time $t = 0$. At this instant we have

$$u(x) = u_0 \cos \left(\frac{-2\pi v x}{v}\right) = u_0 \cos \left(\frac{2\pi v x}{v}\right)$$

Since $v/v = 1/\lambda$, this can equally well be written

$$u(x) = u_0 \cos\left(\frac{2\pi x}{\lambda}\right)$$

The kinetic-energy density is thus given by

$$\frac{dK}{dx} = \tfrac{1}{2}\mu u^2 = \tfrac{1}{2}\mu u_0{}^2 \cos^2\left(\frac{2\pi x}{\lambda}\right)$$

The total kinetic energy in the segment of string between $x = 0$ and $x = \lambda$ is thus

$$K = \tfrac{1}{2}\mu u_0{}^2 \int_0^\lambda \cos^2\left(\frac{2\pi x}{\lambda}\right) dx$$

i.e.,

$$K = \tfrac{1}{4}(\lambda\mu)u_0{}^2 \tag{7-34}$$

This, then, is the kinetic energy associated with one complete wavelength of the disturbance. (You can easily verify that the same answer is obtained by integrating the kinetic-energy density between any two values of x separated by λ at a given instant.)

The potential energy over the same portion of the string must, as we have already seen, be equal to the kinetic energy. For the sake of being quite explicit, however, we will carry out the calculation. From Eq. (7–33) we have

$$\frac{\partial y}{\partial x} = -\frac{2\pi v A}{v} \cos 2\pi v \left(t - \frac{x}{v}\right)$$

Thus at $t = 0$ we have

$$\left(\frac{\partial y}{\partial x}\right)_{t=0} = -\frac{2\pi v A}{v} \cos\left(\frac{2\pi v x}{v}\right) = -\frac{2\pi A}{\lambda} \cos\left(\frac{2\pi x}{\lambda}\right)$$

Hence the potential-energy density [Eq. (7–32)] is given by

$$\frac{dU}{dx} = \frac{2\pi^2 A^2 T}{\lambda^2} \cos^2\left(\frac{2\pi x}{\lambda}\right)$$

Integrating over one wavelength then gives

$$U = \frac{\pi^2 A^2 T}{\lambda}$$

Putting $T = \mu v^2 = \mu v^2 \lambda^2$, this gives us

$$U = \pi^2 A^2 \mu v^2 \lambda \tag{7-35}$$

which can be recognized as equal in magnitude to K, as given by Eq. (7–34), if we use the identity $u_0 = 2\pi v A$.

The total energy per wavelength, E, can be written

$$E = \tfrac{1}{2}(\lambda\mu)u_0{}^2 \tag{7-36}$$

and is thus equal to the kinetic energy that a piece of the string of length λ would have if all of it were moving with the maximum transverse velocity u_0 associated with the wave.

Although we have chosen to make this calculation for a sinusoidal wave, equivalent results can be calculated for other kinds of waveform (see Problem 7–23 for one such example).

THE TRANSPORT OF ENERGY BY A WAVE

Imagine that one end of a very long string is being oscillated transversely so as to generate a sinusoidal wave traveling out along the string. The calculations of the previous section clearly require that this process must involve a continuing input of energy. For each new length λ of the string that is set in motion by the wave, the amount of energy given by Eq. (7–36) must be supplied. The work equivalent to this energy must therefore be supplied by the driving agent (the source) at the end of the string. Let us see how this can be verified.

We shall take the same sinusoidal wave equation as in the last section [Eq. (7–33)]:

$$y(x, t) = A \sin 2\pi\nu \left(t - \frac{x}{v} \right)$$

We shall assume that the string has one end at $x = 0$ and is driven at this point (Fig. 7–19). The driving force, F, equal in magnitude to the tension, T, must be applied in a direction tangent to the string, as shown in the figure. The motion of the end point, assumed to be purely transverse, is given by the equation

$$y_0(t) = A \sin 2\pi\nu t$$

The component of F in the direction of this transverse motion is given by

$$F_y = -T \sin \theta \approx -T \left(\frac{\partial y}{\partial x} \right)_{x=0}$$

From Eq. (7–33) we have

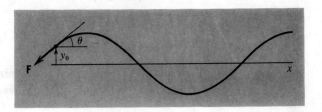

Fig. 7–19 Generation of sinusoidal wave on stretched string, showing applied force vector at an arbitrary instant.

$$\frac{\partial y}{\partial x} = -\frac{2\pi \nu A}{v} \cos 2\pi \nu \left(t - \frac{x}{v} \right)$$

Therefore,

$$F_y = \frac{2\pi \nu A T}{v} \cos 2\pi \nu t$$

We can now calculate the work done in any given time as the integral of $F_y\, dy_0$:

$$W = \int F_y\, dy_0 = \frac{2\pi \nu A T}{v} \int \cos 2\pi \nu t\; d(A \sin 2\pi \nu t)$$

$$= \frac{(2\pi \nu A)^2 T}{v} \int \cos^2 2\pi \nu t\, dt$$

We can express this more simply by recognizing that $2\pi \nu A$ is the maximum speed u_0 of the transverse motion. Thus we can put

$$W = \frac{u_0{}^2 T}{v} \int \cos^2 2\pi \nu t\, dt$$

Let us evaluate this work for one complete period of the wave, by taking the integral from $t = 0$ to $t = 1/\nu$. Then we have

$$W_{\mathrm{cycle}} = \frac{u_0{}^2 T}{2v} \int_0^{1/\nu} (1 + \cos 4\pi \nu t)\, dt$$

The term $\cos 4\pi \nu t$ contributes nothing in this complete cycle, so we have

$$W_{\mathrm{cycle}} = \frac{u_0{}^2 T}{2v\nu} \tag{7–37}$$

Since $T = \mu v^2$ and $\nu = v/\lambda$, this can be expressed in the alternative forms

$$W_{\mathrm{cycle}} = \tfrac{1}{2}(\lambda\mu)u_0{}^2 = 2\pi^2 \nu^2 A^2 \lambda \mu \tag{7–38}$$

which are just twice the values of the kinetic energy and potential energy per wavelength, as given in Eqs. (7–34) and (7–35).

The *rate* of doing work, as described by the mean power input P, is obtained by taking Eq. (7–37) for the work per cycle and multiplying by the number of cycles per unit time (ν). This gives us

$$P = \frac{u_0{}^2 T}{2v} = \tfrac{1}{2}\mu u_0{}^2 v \tag{7–39}$$

(Recall that $T = \mu v^2$.) We recognize P as being equal to the total energy per unit length that the wave adds to the string ($\tfrac{1}{2}\mu u_0{}^2$)

multiplied by the wave speed (v), which may be thought of (at least until the wave reaches the far end of the string) as representing the additional length of string per unit time to become involved in the disturbance. The energy is not retained at the source; it flows along the string, which thus acts as a medium for the transport of energy from one point to another, the speed of transport being equal to the wave speed v.[1] (Note that once a given portion of the string has become fully involved in the wave motion, its average energy remains constant.)

MOMENTUM FLOW AND MECHANICAL RADIATION PRESSURE

It is natural to expect that, associated with the transport of energy by a mechanical wave, there must also be a transport of momentum. And it is tempting to suppose that the ratio of energy transport to momentum transport is essentially the wave speed v (in much the same way as the ratio of energy to momentum for a particle is essentially—but for a factor of $\frac{1}{2}$—equal to the particle speed). This, however, is not, in general, the case. The calculation of the wave momentum involves a detailed consideration of the properties of the medium, and the results can be surprising. For example, one would conclude that the longitudinal waves in a bar that obeys Hooke's law exactly can carry no momentum at all. The perfectly elastic medium in this sense does not exist, but the calculation of the momentum flow in a real medium then becomes a subtle and sometimes difficult matter.

A question closely related to that of momentum flow is the mechanical force exerted by waves on an object that absorbs or reflects them. It is well established, for example, that longitudinal waves in a gas (sound waves) exert a pressure on a surface placed in their path, and the existence of this pressure must certainly be associated with a transport of momentum by the waves. In this particular case the force exerted on a surface by the waves is indeed given in order of magnitude by the rate of energy flow divided by the wave speed—a relation that holds exactly for electromagnetic waves. Once again it should, however, be emphasized that the precise result depends on assumptions about the equation of state (i.e., the equation that relates changes of stress

[1]We are here assuming no dispersion. If the medium is dispersive, it turns out that it is the group velocity that characterizes the velocity of transport of energy.

and density) for the medium. The existence of momentum flow, and of associated longitudinal forces, depends essentially on non-linearities in the equations of motion which are not compatible with strictly sinusoidal wave solutions. This puts the problem outside the scope of our present discussions, so we shall not pursue it further.[1]

WAVES IN TWO AND THREE DIMENSIONS

In Chapter 6 we gave some examples of the normal modes of systems that were essentially two-dimensional—soap films and thin flat plates. The simplest case is that of a membrane (of which a soap film is, in fact, a good example) subjected to a uniform tension S (per unit length) as measured across any line in its plane. If we introduce rectangular coordinates x, y in the plane of the membrane, and describe transverse displacements in terms of a third coordinate, z, then, as we saw, the following wave equation results:

$$\frac{\partial^2 z}{\partial x^2} + \frac{\partial^2 z}{\partial y^2} = \frac{1}{v^2}\frac{\partial^2 z}{\partial t^2} \qquad (7\text{--}40)$$

The wave velocity v is given by

$$v^2 = \frac{S}{\sigma}$$

where σ is the surface density (i.e., mass per unit area) of the membrane.

If the symmetry of such a system is rectangular, it is possible to apply Eq. (7–40) at once and obtain solutions in the form of straight waves, of the form

$$z(x, y, t) = f(\alpha x + \beta y - vt)$$

Suitable superpositions of such waves, in a system with rectangular boundaries, correspond to normal modes such as those shown in Fig. 6–11.

If, on the other hand, the natural symmetry of the system is circular—as it might be, for example, if waves were generated on a membrane by setting one point of it into transverse motion, then it is appropriate to introduce plane polar coordinates r, θ

[1]For fuller discussions of wave momentum and pressure, see the article "Radiation Pressure in a Sound Wave," by R. T. Beyer, *Am. J. Phys.*, **18**, 25 (1950), and the book by R. B. Lindsay, *Mechanical Radiation*, McGraw-Hill, New York, 1960.

in the place of x and y. Let us limit ourselves to a completely symmetrical case, in which the displacement z is independent of θ at a given value of r. Then Eq. (7–40) goes over into the following form:

$$\text{(cylindrical symmetry)} \quad \frac{\partial^2 z}{\partial r^2} + \frac{1}{r}\frac{\partial z}{\partial r} = \frac{1}{v^2}\frac{\partial^2 z}{\partial t^2} \tag{7–41}$$

The traveling waves that represent solutions of this equation are expanding circular wavefronts. One can recognize more or less intuitively that the amplitude of vibration becomes less as r increases, because the disturbance is being spread over the perimeters of circles of increasing radius. The precise solutions are obtained in terms of special functions called Bessel's functions. At sufficiently large r the second term on the right in Eq. (7–41) becomes almost negligible compared to the first, and to some approximation the equation reverts to that for straight wavefronts of constant amplitude. (More accurately, the amplitude falls off approximately as $1/\sqrt{r}$.) This is, of course, the impression one has if one is very far from the origin of circular waves and sees only a small portion of the perimeter of the wavefront.

Finally, we can set up a wave equation for a three-dimensional medium, such as a block of elastic solid, or air not confined to a tube. This also we quoted in Chapter 6:

$$\frac{\partial^2 \Psi}{\partial x^2} + \frac{\partial^2 \Psi}{\partial y^2} + \frac{\partial^2 \Psi}{\partial z^2} = \frac{1}{v^2}\frac{\partial^2 \Psi}{\partial t^2} \tag{7–42}$$

where Ψ is some variable such as the local magnitude of the pressure. The combination of differential operators on the left-hand side is named the Laplacian (after P. S. de Laplace, a near contemporary of Lagrange) and is given the special symbol ∇^2 for short (pronounced "del-squared"). Thus we write Eq. (7–42) in the alternative form

$$\nabla^2 \Psi = \frac{1}{v^2}\frac{\partial^2 \Psi}{\partial t^2} \tag{7–43}$$

As with the two-dimensional medium, if we have a system with rectangular symmetries it is appropriate to look for plane-wave solutions of the wave equation:

$$\Psi(x, y, z, t) = f(\alpha x + \beta y + \gamma z - vt)$$

But, on the other hand, if spherical symmetry suggests itself—as with the waves that would be generated if a small explosion took place deep in the ground—then we introduce the radius r and two angles to define the position of a point. For a system in which the wave amplitude depends on r only, the differential equation reduces to the following:

$$\text{(spherical symmetry)} \quad \frac{\partial^2 \Psi}{\partial r^2} + \frac{2}{r}\frac{\partial \Psi}{\partial r} = \frac{1}{v^2}\frac{\partial^2 \Psi}{\partial t^2} \qquad (7\text{--}44)$$

It is easy to verify that this equation is satisfied by simple harmonic waves whose amplitude falls off inversely with r:

$$\Psi(r, t) = \frac{C}{r}\sin 2\pi(vt - kr) \qquad (7\text{--}45)$$

Remembering that the energy flow for a one-dimensional wave is proportional to the amplitude squared, one can see in Eq. (7–45) the implication that the time average of $[\Psi(r, t)]^2$, multiplied by the area $4\pi r^2$ of a sphere of radius r, defines a rate of outflow of energy that is independent of the distance from a point source that generates the waves. In the absence of dissipation or absorption, this is just what one would expect to find.

PROBLEMS

7–1 Satisfy yourself that the following equations can all be used to describe the same progressive wave:

$y = A \sin 2\pi(x - vt)/\lambda$
$y = A \sin 2\pi(kx - vt)$
$y = A \sin 2\pi[(x/\lambda) - (t/T)]$
$y = -A \sin \omega(t - x/v)$
$y = A \operatorname{Im}\{\exp[j2\pi(kx - vt)]\}$

7–2 The equation of a transverse wave traveling along a string is given by $y = 0.3 \sin \pi(0.5x - 50t)$, where y and x are in centimeters and t is in seconds.

(a) Find the amplitude, wavelength, wave number, frequency, period, and velocity of the wave.

(b) Find the maximum transverse speed of any particle in the string.

7–3 What is the equation for a longitudinal wave traveling in the negative x direction with amplitude 0.003 m, frequency 5 sec^{-1}, and speed 3000 m/sec?

7–4 A wave of frequency 20 sec^{-1} has a velocity of 80 m/sec.

(a) How far apart are two points whose displacements are 30° apart in phase?

(b) At a given point, what is the phase difference between two displacements occurring at times separated by 0.01 sec?

7–5 A long uniform string of mass density 0.1 kg/m is stretched with a force of 50 N. One end of the string ($x = 0$) is oscillated transversely (sinusoidally) with an amplitude of 0.02 m and a period of 0.1 sec, so that traveling waves in the $+x$ direction are set up.

(a) What is the velocity of the waves?

(b) What is their wavelength?

(c) If at the driving end ($x = 0$) the displacement (y) at $t = 0$ is 0.01 m with dy/dt negative, what is the equation of the traveling waves?

7-6 It is observed that a pulse requires 0.1 sec to travel from one end to the other of a long string. The tension in the string is provided by passing the string over a pulley to a weight which has 100 times the mass of the string.

(a) What is the length of the string?

(b) What is the equation of the third normal mode?

7-7 A very long string of the same tension and mass per unit length as that in Problem 7-6 has a traveling wave set up in it with the following equation:

$$y(x, t) = 0.02 \sin \pi(x - vt)$$

where x and y are in meters, t in seconds, and v is the wave velocity (which you can calculate). Find the transverse displacement and velocity of the string at the point $x = 5$ m at the time $t = 0.1$ sec.

7-8 Two points on a string are observed as a traveling wave passes them. The points are at $x_1 = 0$ and $x_2 = 1$ m. The transverse motions of the two points are found to be as follows:

$$y_1 = 0.2 \sin 3\pi t$$
$$y_2 = 0.2 \sin(3\pi t + \pi/8)$$

(a) What is the frequency in hertz?

(b) What is the wavelength?

(c) With what speed does the wave travel?

(d) Which way is the wave traveling? Show how you reach this conclusion.

(*Warning!* Consider carefully if there are any ambiguities allowed by the limited amount of information given.)

7-9 A symmetrical triangular pulse of maximum height 0.4 m and total length 1.0 m is moving in the positive x direction on a string on which the wave speed is 24 m/sec. At $t = 0$ the pulse is entirely located between $x = 0$ and $x = 1$ m. Draw a graph of the transverse velocity versus time at $x = x_2 = +1$ m.

7-10 The end ($x = 0$) of a stretched string is moved transversely with a constant speed of 0.5 m/sec for 0.1 sec (beginning at $t = 0$) and is returned to its normal position during the next 0.1 sec, again at constant speed. The resulting wave pulse moves at a speed of 4 m/sec.

(a) Sketch the appearance of the string at $t = 0.4$ sec and at $t = 0.5$ sec.

(b) Draw a graph of transverse velocity against x at $t = 0.4$ sec.

7–11 Suppose that a traveling wave pulse is described by the equation

$$y(x, t) = \frac{b^3}{b^2 + (x - vt)^2}$$

with $b = 5$ cm and $v = 2.5$ cm/sec. Draw the profile of the pulse as it would appear at $t = 0$ and $t = 0.2$ sec. By direct subtraction of ordinates of these two curves, obtain an appropriate picture of the transverse velocity as a function of x at $t = 0.1$ sec. Compare with what you obtain by calculating $\partial y / \partial t$ at an arbitrary t and then putting $t = 0.1$ sec.

7–12 The figure shows a pulse on a string of length 100 m with fixed ends. The pulse is traveling to the right without any change of shape, at a speed of 40 m/sec.

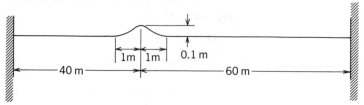

(a) Make a clear sketch showing how the transverse velocity of the string varies with distance along the string at the instant when the pulse is in the position shown.

(b) What is the maximum transverse velocity of the string (approximately)?

(c) If the total mass of the string is 2 kg, what is the tension T in it?

(d) Write an equation for $y(x, t)$ that numerically describes sinusoidal waves of wavelength 5 m and amplitude 0.2 m traveling to the left (i.e., in the negative x direction) on a very long string made of the same material and under the same tension as above.

7–13 A pulse traveling along a stretched string is described by the following equation:

$$y(x, t) = \frac{b^3}{b^2 + (2x - ut)^2}$$

(a) Sketch the graph of y against x for $t = 0$.

(b) What are the speed of the pulse and its direction of travel?

(c) The transverse velocity of a given point of the string is defined by

$$v_y = \frac{\partial y}{\partial t}$$

Calculate v_y as a function of x for the instant $t = 0$, and show by means of a sketch what this tells us about the motion of the pulse during a short time Δt.

Kink

7–14 A closed loop of uniform string is rotated rapidly at some constant angular velocity ω. The mass of the string is M and the radius is R. A tension T is set up circumferentially in the string as a result of its rotation.

(a) By considering the instantaneous centripetal acceleration of a small segment of the string, show that the tension must be equal to $M\omega^2 R/2\pi$.

(b) The string is suddenly deformed at some point, causing a kink to appear in it, as shown in the diagram. Show that this could produce a distortion of the string that remains stationary with respect to the laboratory, regardless of the particular values of M, ω, and R. But is this the whole story? (Remember that pulses on a string may travel both ways.)

7–15 Two identical pulses of equal but opposite amplitudes approach each other as they propagate on a string. At $t = 0$ they are as shown in the figure. Sketch *to scale* the string, and the velocity profile of the string mass elements, at $t = 1$ sec, $t = 1.5$ sec, $t = 2$ sec.

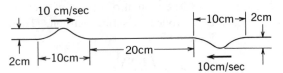

7–16 It is desired to study the rather rapid vertical motion of the moving contact of a magnetically operated switch. To do this, the contact is attached to one end (O) of a horizontal fishline of total mass 5 g (5×10^{-3} kg) and total length 12.5 m. The other end of the line passes over a small, effectively frictionless pulley, and a mass of 10 kg is hung from it, as shown in the sketch. The contact is actuated so that the switch (initially open) goes into the closed position, remains closed for a short time, and opens again. Shortly thereafter the string

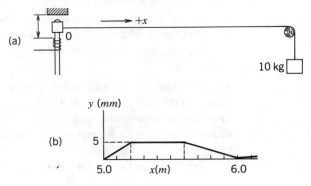

is photographed, using a high-speed flash, and it is found to be deformed between 5 and 6 m, as shown ($x = 0$ is the point O where the string is connected to the contact.)

(a) For how long was the switch *completely closed*?

(b) Draw a graph of the displacement of the contact *as a function of time*, taking $t = 0$ to be the instant at which the contact first began to move.

(c) What was the maximum speed of the contact? Did it occur during closing or during opening of the switch?

(d) At what value of t was the photograph taken?
(Assume $g = 10 \text{ m/sec}^2$.)

7–17 The following two waves in a medium are superposed:

$$y_1 = A \sin(5x - 10t)$$
$$y_2 = A \sin(4x - 9t)$$

where x is in meters and t in seconds.

(a) Write an equation for the combined disturbance.

(b) What is its group velocity?

(c) What is the distance between points of zero amplitude in the combined disturbance?

7–18 The motion of ripples of short wavelength ($\lesssim 1$ cm) on water is controlled by surface tension. The phase velocity of such ripples is given by

$$v_p = \left(\frac{2\pi S}{\rho\lambda}\right)^{1/2}$$

where S is the surface tension and ρ the density of water.

(a) Show that the group velocity for a disturbance made up of wavelengths close to a given λ is equal to $3v_p/2$.

(b) What does this imply about the observed motion of a group of ripples traveling over a water surface?

(c) If the group consists of just two waves, of wavelengths 0.99 and 1.01 cm, what is the distance between crests of the group?

7–19 The relation between frequency v and wave number k for waves in a certain medium is as shown in the graph. Make a qualitative statement (and explain the basis for it) about the relative magnitudes of the group and phase velocities at any wavelength in the range represented.

7–20 Consider a U-tube of uniform cross section with two vertical arms. Let the total length of the liquid column be l. Imagine the liquid to be oscillating back and forth, so that at any instant the levels in the side arms are at $\pm y$ with respect to the equilibrium level, and all the liquid has the speed dy/dt.

(a) Write down an expression for the potential energy plus

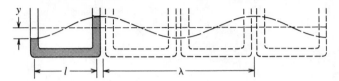

kinetic energy of the liquid, and hence show that the period of oscillation is $\pi\sqrt{2l/g}$.

(b) Imagine that a succession of such tubes can be used to define a succession of crests and troughs as in a water wave (see the diagram). Taking the result of (a), and the condition $\lambda \approx 2l$ implied by this analogy, deduce that the speed of waves on water is something like $(g\lambda)^{1/2}/\pi$. (Assume that only a small fraction of the liquid is in the vertical arms of the U-tube.)

(c) Use the exact result, $v = (g\lambda/2\pi)^{1/2}$, to calculate the speed of waves of wavelength 500 m in the ocean.

7–21 Consider a system of N coupled oscillators ($N \gg 1$), each separated from its nearest neighbors by a distance l.

(a) Find the wavelength and frequency of the nth mode of oscillation.

(b) Find the phase and group velocities for this mode. What are they for the cases $n \ll N$ and $n = N + 1$?

7–22 You are given the problem of analyzing the dynamics of a line of cars moving on a one-lane highway. One approach to this problem is to assume that the line of cars behaves like a group of coupled oscillators. How would you set this problem up in a tractable way? Make lots of assumptions.

7–23 One end of a stretched string is moved transversely at constant velocity u_y for a time τ, and is moved back to its starting point with velocity $-u_y$ during the next interval τ. As a result, a triangular pulse is set up on the string and moves along it with speed v. Calculate the kinetic and potential energies associated with the pulse, and show that their sum is equal to the total work done by the transverse force that has to be applied at the end of the string.

7–24 Consider a longitudinal sinusoidal wave $\xi = \xi_0 \cos 2\pi k(x - vt)$ traveling down a rod of mass density ρ, cross-sectional area S, and Young's modulus Y. Show that if the stress in the rod is due solely to the presence of the wave, the kinetic-energy density is $\frac{1}{2}\rho S(\partial\xi/\partial t)^2$, and the potential-energy density is $\frac{1}{2}YS(\partial\xi/\partial x)^2$. Thus show that the kinetic energy per wavelength and the potential energy per wavelength both equal $\frac{1}{4}(\rho S\lambda)u_0^2$, where u_0 is the maximum particle velocity $(\partial\xi/\partial t)$.

7–25 Verify that the wave equation for spherically symmetric waves [Eq. (7–44)] is satisfied by simple harmonic waves whose amplitudes fall off inversely with r.

251   Problems

We shall see in this chapter how sounds quarrel,
fight, and when they are of equal strength destroy
one another, and give place to silence.

ROBERT BALL, *Wonders of Acoustics* (1867)

8

Boundary effects and interference

THE PRECEDING CHAPTER was concerned with waves that could be imagined as traveling uninterrupted in a specified medium. This chapter is chiefly about some of the effects that take place when a traveling wave encounters a barrier, or a different medium, or small obstacles. Such effects represent an enormous field of study, and the present account is not intended to be more than a first glimpse of the analysis of these phenomena. We shall begin with our old standby, the stretched string, and will consider what happens when a traveling wave on a string encounters a discontinuity of some kind.

REFLECTION OF WAVE PULSES

In discussing the connection between standing waves and traveling waves on a stretched string, we necessarily made some reference to the conditions that exist at the two ends of any string of limited length. We pointed out that, as a matter of experience, one can set up a standing wave by agitating one end of a string, thereby generating a traveling wave which undergoes some process of reflection at the far end. The outgoing and returning waves then conspire to produce a standing-wave pattern with nodes at fixed positions.

More quantitatively, we recognized that a given normal mode on a string with fixed ends can be regarded as the superposition of two sine waves of equal amplitude, wavelength, and frequency, traveling in opposite directions. To be specific, we noted that the following two statements are mathematically equivalent:

Normal mode:

$$y(x, t) = A \sin \left(\frac{n\pi x}{L} \right) \cos \omega t$$

Two traveling waves:

$$y(x, t) = \frac{A}{2} \sin \left(\frac{n\pi x}{L} - \omega t \right) + \frac{A}{2} \sin \left(\frac{n\pi x}{L} + \omega t \right)$$

If we take the second of these statements, and fix attention on the conditions at $x = 0$ or $x = L$, we have

$$y(0, t) = y(L, t) = \frac{A}{2} \sin(-\omega t) + \frac{A}{2} \sin \omega t$$

$$= -\frac{A}{2} \sin \omega t + \frac{A}{2} \sin \omega t$$

What this says is that these oppositely traveling waves must, at all times, produce equal and opposite displacements at the fixed ends—which is, of course, pretty obviously necessary. And the main point is that this same condition must define the reflection process for any traveling wave when it encounters a rigid boundary.

Let us take a second look at another such superposition process. In connection with Fig. 7–15, we discussed the superposition of two symmetrical pulses of opposite displacements, traveling in opposite directions along a string. An interesting fact can be noted in this example: The point on the string at which the two pulses meet remains at rest at all times! The waves pass through in opposite directions without causing any displacement of the point at any time. We could consider the point to be rigidly fixed to a wall without altering the wave pattern in any manner. This gives us the clue as to what happens when a wave pulse is incident upon the end of a string which is held stationary: A pulse of opposite displacement is reflected from the end and travels back toward the source.

This inverted reflection is not so mysterious when we consider that the arrival of a positive displacement will exert an upward force on the support which holds the end fixed (see Fig. 8–1). By Newton's third law, the support exerts a reaction

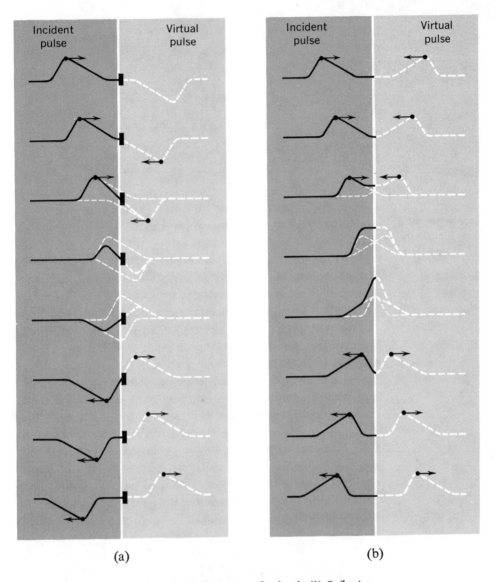

(a)   (b)

Fig. 8-1 (a) Reflection at a fixed end. (b) Reflection
at a free end. One may think of reflection as if the
string extended indefinitely beyond its actual terminus.
The pulse can be considered to continue on into the
imaginary portion as though the support were not there,
while at the same time a "virtual" pulse, which has
been traveling in the imaginary portion, moves out into
the real string and forms the reflected pulse. The
nature of the reflected pulse depends on whether the end
is fixed or free. (Figure adapted from F. W. Sears
and M. W. Zemansky, University Physics, 3rd ed.,
Part I, Addison-Wesley, Reading, Mass., 1963.)

255   Reflection of wave pulses

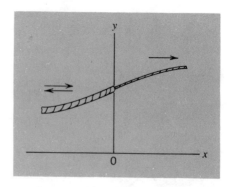

Fig. 8–2 Reflection and transmission at a junction between dissimilar strings.

force in the opposite direction back on the string, thus generating a pulse of *opposite* polarity which travels backward toward the source.

If the end of the string is completely free to move (e.g., if it were tied to a massless ring on a frictionless vertical rod[1]), the arrival of a positive pulse will exert an upward reaction force back on the string, generating a pulse of positive polarity. This positive pulse is then transmitted back along the string. Reflection from a "free" end thus produces a pulse of the *same* polarity traveling back toward the source.

If a string with a certain tension and mass per unit length is fastened to another string with a different μ, in general some reflection will take place (as well as some transmission) at the discontinuity. To see quantitatively how this comes about, consider a pulse of the form $f_1(t - x/v_1)$ moving along a stretched cord of linear density μ_1, which is joined to a cord of linear density μ_2 at $x = 0$ (Fig. 8–2).[2] Assuming partial reflection and partial transmission at the junction, the transverse displacements in the two strings can be assumed to be given by the following equations:

$$y_1(x, t) = f_1\left(t - \frac{x}{v_1}\right) + g\left(t + \frac{x}{v_1}\right)$$

$$y_2(x, t) = f_2\left(t - \frac{x}{v_2}\right)$$

(8–1)

Because the ends of the two cords remain in contact with each

[1] Another method of achieving a "free" termination is to tie it to a very much less massive string.

[2] Writing the pulse as $f(t - x/v)$ instead of our usual $f(x - vt)$ is more appropriate in this analysis, basically because when a wave passes from one medium to another the wavelength changes but the frequency does not. Thus we associate the changed factors with the x coordinate and not with t.

other, the transverse displacements, y, at the point $x = 0$, must be the same for both cords. Also, at each instant the cords must join with equal slopes and have equal tensions; otherwise the element of mass represented by the junction would be given a very large acceleration. Thus we have the following two conditions:

$$y_1(0, t) = y_2(0, t)$$

$$\frac{\partial y_1}{\partial x}(0, t) = \frac{\partial y_2}{\partial x}(0, t)$$

i.e.,

$$f_1(t) + g_1(t) = f_2(t) \tag{8-2}$$

$$\frac{1}{v_1} f_1'(t) - \frac{1}{v_1} g_1'(t) = \frac{1}{v_2} f_2'(t) \tag{8-3}$$

Integrating Eq. (8–3), we have

$$v_2 f_1(t) - v_2 g_1(t) = v_1 f_2(t) \tag{8-4}$$

Solving Eqs. (8–2) and (8–4) for g_1 and f_2 in terms of f_1, we find

$$g_1(t) = \frac{v_2 - v_1}{v_2 + v_1} f_1(t)$$

$$f_2(t) = \frac{2v_2}{v_2 + v_1} f_1(t) \tag{8-5}$$

As they stand, equations (8–5) are merely a description of the state of affairs at $x = 0$ at any arbitrary value of t. Now, however, we shall introduce a somewhat subtle but very important piece of reasoning. What equations (8–5) do is to relate the values of f_1, g_1, and f_2 at the same value of their argument. For any given value, say τ, of this argument, we have $g_1(\tau) = $ const. $\times f_1(\tau)$, and $f_2(\tau) = $ const. $\times f_1(\tau)$. But õne is not restricted to interpreting τ as the value of t at $x = 0$. It can be used to define all other values of x and t that are related in the manner required by the basic statement of a given traveling wave. Thus the function f_1 is *defined* to be a function of the argument $t - x/v_1$, and the function g_1 is *defined* to be a function of the argument $t + x/v_1$. Suppose each of these arguments is set equal to the same value τ, as required by Eq. (8–5). Clearly we cannot use the same pair of values of x and t for both; let us therefore label the values as x_f, t_f, and x_g, t_g. Then we have

$$\tau = t_f - \frac{x_f}{v_1} = t_g + \frac{x_g}{v_1}$$

If we put $t_f = t_g = t$, then we must have

$$x_g = -x_f$$

and what *this* means is that the displacement associated with pulse g_1 at any given instant, at any given value of x, is directly related to the value of f_1 as calculated at the *same* time at the position $-x$. Specifically, according to the first of equations (8–5), we have

$$g_1\left(t + \frac{-x}{v_1}\right) = \frac{v_2 - v_1}{v_2 + v_1} f_1\left(t - \frac{x}{v_1}\right) \tag{8–6a}$$

And what this says is that the reflected pulse, besides being scaled down by the factor $(v_2 - v_1)/(v_2 + v_1)$, is reversed right to left with respect to the incident pulse. If $v_2 < v_1$, it is also turned upside down.

In a similar way we can relate the transmitted waveform f_2 to the incident waveform f_1. At a given value of t the corresponding values of x (call them x_1 and x_2) are defined by the relation

$$\tau = t - \frac{x_1}{v_1} = t - \frac{x_2}{v_2}$$

Hence $x_2 = (v_2/v_1)x_1$, and the second of equations (8–5) requires us to put

$$f_2\left(t - \frac{v_2 x/v_1}{v_2}\right) = \frac{2v_2}{v_2 + v_1} f_1\left(t - \frac{x}{v_1}\right) \tag{8–6b}$$

This tells us that, compared to the incident pulse, the transmitted

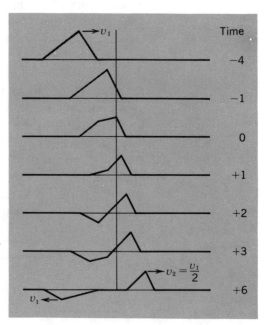

Fig. 8–3 *Partial reflection and transmission of triangular wave pulse at the junction of two strings. Pulse is incident from string with higher wave speed.* $(v_2 = \frac{1}{2}v_1.)$

pulse suffers not only a change in height but also a scale change along x.

In using the above relationships, it is to be noted that if the pulse f_1 is incident from the negative x direction and if the junction is at $x = 0$, then the functions f_1 and g_1 represent physically real displacements only if $x \leq 0$, whereas f_2 represents a physically real displacement only if $x \geq 0$. Thus, for example, in using Eq. (8–6a), we find the real displacement in the reflected pulse g_1, at some negative value of x, by considering what the displacement of the incident pulse f_1 *would* have been if it had continued on into the region of positive x, and then multiplying by the factor $(v_2 - v_1)/(v_2 + v_1)$. In Fig. 8–3 we show the development of the reflected and transmitted pulses from a given incident pulse for the particular case $v_2 = v_1/2$.

As extreme cases of Eq. (8–6a) we have the following:

a. String 2 infinitely massive:

$v_2 = 0$

$$g\left(t + \frac{-x}{v_1}\right) = -f_1\left(t - \frac{x}{v_1}\right)$$

b. String 2 massless or absent:

$v_2 = \infty$

$$g\left(t + \frac{-x}{v_1}\right) = f_1\left(t - \frac{x}{v_1}\right)$$

These then represent the two situations shown in Fig. 8–1. Figure 8–4 shows some actual examples of the reflection and transmission of pulses traveling along stretched springs.

IMPEDANCES: NONREFLECTING TERMINATIONS[1]

The kind of behavior discussed in the last section can be treated in a very illuminating way by introducing the concept of the *mechanical impedance* of a physical system subjected to driving forces. This impedance is defined as the ratio of the driving force to the associated velocity of displacement. You will recognize here a strong similarity to the electrical concept of resistance, which is the ratio of an applied voltage to the associated current

[1]This section may be omitted without loss of continuity. (But it is not difficult, and may be quite instructive, given some acquaintance with the properties of basic electric circuit elements.)

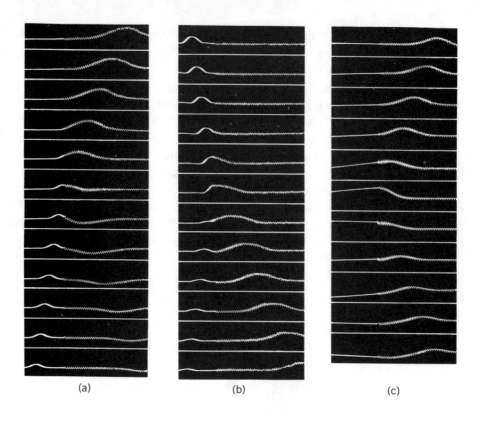

(a)   (b)   (c)

Fig. 8–4   Photographs of pulses encountering the boundary between two media. (a) Pulse passing from a light spring (right) to a heavy spring. At the junction the pulse is partially transmitted and partially reflected. You will note that the reflected pulse is upside down. (b) Pulse passing from a heavy spring (left) to a light spring. At the junction the pulse is partially transmitted and partially reflected. The reflected pulse is right side up. (c) Pulse on a spring reflected from a junction with a very light thread. The whole pulse returns right side up. The blurring of pictures indicates that the particles of the thread are moving at high speed as the pulse passes. Can you determine the direction of this motion in each of the frames? (Photographs from Physical Science Study Committee, Physics, Heath, Boston, 1965.)

(and the current is the rate of flow of charge). But in mechanical and electrical systems alike, there is in general a phase difference between the driving force and the velocity (or between voltage and current). Ohm's law expresses a relation in which voltage and current are always in phase. Thus, for example, if we apply across the ends of a resistor a voltage given by

$$V = V_0 \cos \omega t$$

then the resulting current is given by

$$I = I_0 \cos \omega t$$

where

$$I_0 = \frac{V_0}{R}$$

But if, for example, this same alternating voltage were applied across the plates of a capacitor, it would be the charge q, not the current I, that was in phase with the voltage, for we have

$$q = CV$$

$$I = \frac{dq}{dt}$$

Hence, if

$$V = V_0 \cos \omega t$$

we have

$$I = -\omega C V_0 \sin \omega t$$

i.e.,

$$I = I_0 \cos \left(\omega t + \frac{\pi}{2} \right)$$

where

$$I_0 = \omega C V_0$$

There is thus a phase difference of 90° between V and I in this case. And if one connected the resistor and the capacitor in series, with the voltage across the combination, then the phase difference would be neither 0° nor 90°. In these more general situations, therefore, the ratio of driving voltage to current involves both a magnitude and a phase, and the quantity which embodies the specification of both of these is called the *impedance* of the system. You will recall that the relation between driving force and displacement in a mechanical oscillator with damping (Chapter 4) was very much of this same character, and, just as in that case, the use of complex quantities provides a simple and economical way of displaying both the amplitude and the phase relationships. It is customary, in fact, to characterize the impedance by a single complex quantity, Z, and to express voltage and current by complex exponentials. Thus, to define the impedance of a capacitor, we would put

$$V = V_0 e^{j\omega t}$$

$$q = C V_0 e^{j\omega t}$$

$$I = \frac{dq}{dt} = j\omega C V_0 e^{j\omega t} = j\omega C V$$

Therefore,

$$Z_C = \frac{V}{I} = \frac{1}{j\omega C}$$

Returning now to the matter of traveling waves on a string, consider first the generation of such a wave through the application of a transverse driving force at any point. Assume that the string lies to the right of this point so that the wave being generated is of the form

$$y(x, t) = f(t - x/v)$$

Then we have

$$v_y = \frac{\partial y}{\partial t} = f'\left(t - \frac{x}{v}\right)$$

$$F_y = -T\frac{\partial y}{\partial x} = \frac{T}{v}f'\left(t - \frac{x}{v}\right)$$

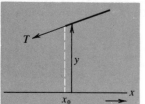

Fig. 8-5 Conditions at the driven end of a stretched string.

(Note that the force exerted *on* the string is the negative of $T\,\partial y/\partial x$—see Fig. 8-5.)

We thus define the impedance Z through the equation

$$Z = \frac{F_y}{v_y} = \frac{T}{v} \tag{8-7}$$

Since $v = \sqrt{T/\mu}$, this identifies for us what we can call the *characteristic impedance* of the string in terms of its tension and linear density:

$$Z = (T\mu)^{1/2} \tag{8-8}$$

This quantity is purely real; the driving force and the velocity are always in phase with one another—in electrical terminology, this impedance is purely resistive.

Now let us consider, in these same terms, the conditions at a junction between two different strings. As before, we shall choose $x = 0$ at the junction point, and shall assume a wave incident from negative x along string 1. Thus equations (8-1) again describe the form of solution we expect:

$$y_1(x, t) = f_1\left(t - \frac{x}{v_1}\right) + g_1\left(t + \frac{x}{v_1}\right)$$

$$y_2(x, t) = f_2\left(t - \frac{x}{v_2}\right)$$

At $x = 0$ we thus require

$$v_y = f_1'(t) + g_1'(t) = f_2'(t)$$

$$F_y = \frac{T_1}{v_1}f_1'(t) - \frac{T_1}{v_1}g_1'(t) = \frac{T_2}{v_2}f_2'(t)$$

(8–9)

(In the second equation we require only that the *transverse* forces should be the same. One could, for example, imagine a difference between the magnitudes of the tensions T_1, T_2 if two stretched strings were connected via a ring around a smooth rod, simulating a rigid connection with respect to displacement along x, but offering no resistance along y.)

Introducing the characteristic impedances Z_1, Z_2 of the two strings, we thus have the following two conditions:

$$f_1'(t) + g_1'(t) = f_2'(t)$$
$$Z_1 f_1'(t) - Z_1 g_1'(t) = Z_2 f_2'(t)$$

From those equations we can proceed to results just like Eqs. (8–5) and (8–6), except that now we have

$$g_1(0, t) = \frac{Z_1 - Z_2}{Z_1 + Z_2} f_1(0, t)$$

$$f_2(0, t) = \frac{2Z_1}{Z_1 + Z_2} f_1(0, t)$$

(8–10)

We see that, in these terms, the amount of reflection that occurs when a traveling wave encounters a junction is specified entirely by the characteristic impedance presented to it at the junction. It does not have to be another string, but can be anything at all, characterized by a certain value of F_y/v_y. We again recognize in Eq. (8–10) the two results already discussed: (1) infinite impedance Z_2, giving $g_1(0, t) = -f_1(0, t)$ and (2) zero impedance Z_2, giving $g_1(0, t) = +f_1(0, t)$.

But now let us consider the possibility of *zero reflection*. According to the first of equations (8–10) this is achieved by putting $Z_2 = Z_1$. One way of fulfilling this is to have another string of exactly the same tension and linear density—which, of course, is no junction at all. Another way [Eq. (8–8)] is to have a second string of different tension and linear density, but having $T_2\mu_2 = T_1\mu_1$. *But a third way is to have the end of our first string dipping into a tank of oil of the right consistency.* For we are very familiar with the law of viscous resistance, which for low speeds gives a force proportional to the velocity, and this is just the law we need to define a constant impedance according to the basic definition expressed in equation (8–7). Of course, the proportionality as such is not enough; the actual value of F/v must be equal to the value of $\sqrt{T\mu}$ for the string. But we have here the possibility of an ideal termination for the string. Waves traveling along the string in one direction advance into the oil tank and vanish. By terminating the string in this way, it can be made to behave just as if it were infinitely long; one says that the load, represented by the oil tank, is perfectly matched to the string. All the energy that is carried to the end of the string by the advancing waves is caught and absorbed there. The analogy with the problem of conveying electrical energy from a source to a load as effectively as possible is very apparent, and this matter of correct impedance matching is, of course, of enormous practical importance—another example of the ubiquity of problems that can be related to the behavior of a simple stretched string.

One last remark on this question of junctions. There is no such thing as a completely abrupt transition from one medium to another. There will always be some nonzero distance (even if it is only one atomic diameter) over which the transition occurs. Calculations of the type we have made will describe the situation very well if the length of the transition region is very small compared to the wavelength involved. But if the wavelength is small enough, or the transition gradual enough, one may cease to have any appreciable amount of reflection. An extreme case is an imagined completely smooth variation of properties along the string. For example, consider the uniform string hanging vertically (with $\mu = $ constant but T increasing linearly with distance up from the bottom) as shown in Fig. 5–1; or a uniformly tapered string at constant tension throughout. These have no identifiable discontinuities at which Eq. (8–1) or (8–9) might be applied. An incident wave is led by the nose, as it were. It suffers a smooth change of wavelength and can be brought out at the far end in a very different condition than it had initially. Such carefully graduated systems are frequently used in acoustical and electrical wave propagation.

LONGITUDINAL VERSUS TRANSVERSE WAVES: POLARIZATION

It is perhaps appropriate at this point to comment briefly on the basic types of wave disturbances—transverse and longitudinal—

that we have encountered in the study of one-dimensional wave propagation.

The stretched string is essentially a carrier of transverse waves. A long *spring*, on the other hand, is capable of carrying both transverse and longitudinal disturbances. In this respect a spring is a better analogue of a real solid, which can also carry both transverse (shear) waves and longitudinal (compressional) waves. A column of liquid or gas, in contrast to a solid, has no elastic resistance to change of shape, only to change of density. Thus a column of a fluid (e.g., air) carries only longitudinal waves, except—and it is a very important exception—when gravity or surface tension provides in effect an elastic restoring force against transverse deformations.

With transverse waves, we may need to recognize the possible existence of two different directions of polarization for the vibrations—perpendicular to one another and to the direction of propagation. It may even be that these different polarization states have different wave speeds associated with them—as, for example, in a crystalline medium in which the interatomic spacings are closer in one direction than another. Thus it is quite conceivable that in an anisotropic crystal there may be three different wave speeds along a given direction—one for longitudinal waves and two for the distinct directions of transverse polarization.

When we consider a one-dimensional wave of any kind encountering a boundary or a barrier, the results developed in the last two sections will describe what happens. It may be worth pointing out, however, that a given interface may behave differently with respect to longitudinal and transverse waves. Suppose, for example, that water rests in a tank with smooth vertical walls. The interface between water and wall then acts as an almost completely rigid boundary with respect to longitudinal waves, but as a completely free end with respect to transverse waves. If standing waves were to be set up, the wall would represent a node for longitudinal vibrations of the water but an antinode for transverse vibrations.

WAVES IN TWO DIMENSIONS

At this point we shall take leave of the purely one-dimensional problems, so as to devote some attention to phenomena which, for the most part, require at least a two-dimensional space (e.g., waves on a surface) for their appearance. These are phenomena which involve a change in direction of a traveling wave, or which involve the superposition of disturbances arriving at a given point from different directions. Essentially these same phenomena also occur in the propagation of waves in three dimensions, but the two-dimensional cases are easier to consider and embody most of the important ideas.

Basically, we shall be dealing with various kinds of solutions to the two-dimensional wave equation, as expressed in one or other of the two forms quoted in Chapter 7 [Eqs. (7–40) and (7–41)][1]:

$$\frac{\partial^2 z}{\partial x^2} + \frac{\partial^2 z}{\partial y^2} = \frac{1}{v^2}\frac{\partial^2 z}{\partial t^2}$$

or $$\frac{\partial^2 z}{\partial r^2} + \frac{1}{r}\frac{\partial z}{\partial r} = \frac{1}{v^2}\frac{\partial^2 z}{\partial t^2}$$

(8–11)

For the most part, however, we shall be able to confine our attention to two special forms of wave:

1. Plane waves or straight waves (the latter being a more appropriate description for waves on a surface). Such waves are generated by oscillations of a straight or flat object of linear dimensions large compared to the wavelength.

2. Circular waves, generated by an object whose linear dimensions are small compared to the wavelength. (Such waves in three dimensions would be called spherical.)

As we mentioned in Chapter 7, circular waves at large distance from the source become in effect straight waves, which often simplifies the analysis of their behavior. In particular, at large r we can, to some approximation, ignore the further decrease in amplitude that must, in principle, be taken into account if a further change of r is involved. This simplification applies particularly to the consideration of interference effects due to two or more small sources.

We shall not be concerned with *solving* equations (8–11) in any rigorous sense. Instead, we shall start with the assumption that we have straight waves or circular waves, as the case may be, and will consider their behavior in various physical situations. A complete and accurate solution to any problem in wave propagation would, in principle, mean solving the basic differential equation subject to the restrictions represented by the particular conditions at all boundaries. *Very* few situations can be exactly analyzed in this way, and so one resorts to physically reasonable

[1]Note that the second equation is a special case, based on the assumption that the displacement z is independent of the direction θ.

approximations that in most cases are fully justified by their success. At the root of most of these approximate treatments is a concept that was first introduced by C. Huygens in 1678 and further developed by J. A. Fresnel in 1816 and subsequently. This concept is that, as a wave progresses through a medium, one can treat each point on the advancing wavefront as a new source. The detailed development of this idea is the subject of the next section.

THE HUYGENS–FRESNEL PRINCIPLE

Suppose that a disturbance occurs at some point in a tank of water—a small object is dropped into the water, for example, or the surface is touched with a pencil point. Then an expanding circular wave pulse is created. Ignoring the effects of dispersion, we can say that the pulse expands at some speed v. If the pulse is created at the origin at $t = 0$, then particles of the medium at a distance r from the origin are set in motion at $t = r/v$. It was Huygens' view that the effects occurring at $r + \Delta r$ at time $t + \Delta r/v$ could be ascribed to the agitation of the medium at r at time t, thus treating the disturbance very explicitly as something handed on from point to adjacent point through the medium. Both Huygens and Fresnel applied this idea to the propagation of light, in which the behavior of the medium lay beyond the scope of observation. It is quite probable, however, that this picture of things was suggested by the observed behavior of ripples on water. In particular, if waves traveling outward from a source encounter a barrier with only a tiny aperture in it ("tiny" meaning of width small compared to a wavelength), then this aperture appears to act just like a new point source, from which circular waves spread out. This phenomenon is shown in Fig. 8–6. It does not matter

Fig. 8–6 Generation of Huygens' wavelets at a narrow aperture in an advancing wavefront. (a) Circular primary waves. (From R. W. Pohl, Physical Principles of Mechanics and Acoustics, Blackie, London, 1932.) (b) Straight primary waves. (From the film "Ripple Tank Phenomena" Part II, Education Development Center, Newton, Mass.)

(a)

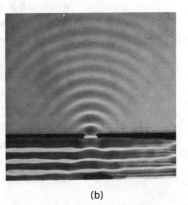
(b)

267   The Huygens-Fresnel principle

whether the original waves are straight or circular; what the small aperture does is to act as a source of circular waves in either case. This is very reasonable, because the effect of the barrier is to suppress all propagation of the original disturbance except through the aperture at which the displacement of the medium is free to communicate itself further.

Huygens' principle accounts nicely for the fact that an unimpeded circular wave pulse gives rise to a subsequent circular wavefront and a straight pulse gives rise to a straight wavefront. Figure 8–7, from Huygens' original book,[1] indicates how, given a circular wavefront HBGI, there will be developed from it at a later time a circular wavefront DCEF. This comes about because each point, such as B, gives rise to a circular wavelet KCL, and the totality of these wavelets generates a reinforcement along the line DCEF that is tangent to them all at a given instant. This locus is characterized by the fact that the shortest distance between it and the original wavefront is everywhere equal to $v \, \Delta t$, where Δt is the time elapsed since the wavefront was at HBGI. A similar construction for a straight wavefront implies that this generates a subsequent wavefront parallel to itself.

There is, however, more in this than meets the eye. The Huygens construction, as we have described it, would define *two* subsequent wavefronts, not one. In addition to a new wavefront farther away from the source, there would be another one corresponding to a wavefront moving back toward the source. But we know that this does not happen. If the Huygens way of

[1]C. Huygens, *Treatise on Light*, 1690 (translated by S. P. Thompson, Dover, New York, 1962).

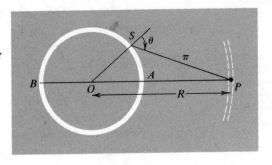

Fig. 8–8   Point S, on an expanding circular wave originating from O, acts as a secondary source whose effect at P depends upon the obliquity angle θ as well as on the distance SP.

visualizing wave propagation is to be acceptable, it must incorporate the unidirectional property of a traveling wave. This can be achieved by requiring that the disturbance starting out from a given point in the medium at a given instant is not equally strong in all directions. Specifically, if O (Fig. 8–8) is the true original source, and S is the origin of a Huygens wavelet, and P is the point at which the disturbance is being recorded, then the effect at P due to the region near S is a function $f(\theta)$ of the angle θ between OS and SP. In particular, $f(\theta) = 0$ for $\theta = \pi$.

As far as our present discussion is concerned, the Huygens construction offers a useful but essentially *qualitative* contribution to the analysis of wave propagation. To do more with it is, in fact, a quite difficult matter. One must define the properties of the secondary sources on an advancing wavefront in such a way that they produce the precise effect that is required of them. A specific mathematical formulation of Huygens' principle in these terms was published by H. Helmholtz in 1859 and was developed further by G. Kirchhoff in 1882.[1] Despite its artificiality, the method is very valuable in the analysis of the optical interference effects that occur when a beam of light is partially interrupted by obstacles. And even without the mathematics, as we shall see, one can use the Huygens approach as a guide.

More often than not, we are dealing with continuing sinusoidal waves, rather than with individual pulses—and even an individual pulse is describable, as we have seen, in terms of superpositions of infinite wave trains, via Fourier analysis. This means

[1]This was for wave propagation in three dimensions. Actually—and surprisingly—the formulation and use of the principle for two dimensions is more difficult and less clearcut than for three. But this is an esoteric point, quite unsuitable for discussion here. For fuller discussions of Huygens' principle, see B. Rossi, *Optics*, Addison-Wesley, Reading, Mass., 1957, or (for a thorough mathematical discussion) B. B. Baker and E. T. Copson, *The Mathematical Theory of Huygens' Principle*, Oxford University Press, New York, 1950.

that we can, at every instant, regard each point on an arbitrarily chosen surface as a secondary source of Huygens wavelets. Although the waves themselves are advancing, the amplitude of the sinusoidal disturbance at any given point is independent of time. This means that, once account has been taken of the *relative* phases of disturbances arriving at a given point from any designated surface, the explicit time dependence can often be disregarded. This will become apparent as we consider specific problems in diffraction and interference.

REFLECTION AND REFRACTION OF PLANE WAVES

Just as with waves on a string, we can, in general, expect a partial reflection and a partial transmission of waves in a medium when they encounter a boundary between two different media. But, with waves in two or three dimensions, we must now also consider the possible changes in direction.

The simplest case is if a straight wave strikes a straight boundary. We then have the familiar laws of reflection and refraction as described in Snell's laws. These results are easily obtained by means of the Huygens construction. In Fig. 8–9, the line AA' represents a straight wavefront at the instant when the point A encounters the boundary. At a later time, the wavefront has advanced, in the original medium, to the position BB'

Fig. 8–9 (a) Reflection and refraction by Huygens' construction. (b) Proof of Snell's Law by Huygens' construction.

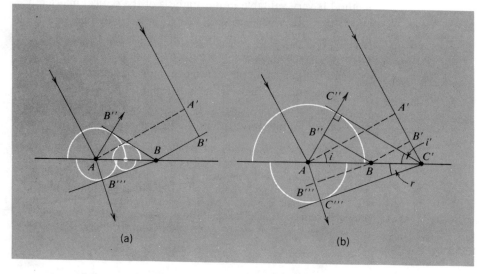

(a)   (b)

270   Boundary effects and interference

[Fig. 8–9(a)]. Each successive point along the boundary between A and B, as it is reached by the wavefront, becomes the center of new Huygens wavelets, advancing into the second medium and traveling back into the original medium. The tangents to these wavelets will be the new wavefronts.

Still later, the original wavefront touches the boundary at point C' [Fig. 8–9(b)]. At that same moment, the wavelet that started out earlier from point A, spreading back into the original medium, will have attained a radius AC''. The line $C'C''$, tangent to this latter wavelet, will also be tangent to all the wavelets arising from the points along the boundary between A and A'.[1] The line $C'C''$ is a new wavefront. And from the geometry of the figure, if i and i' denote the angles made with the boundary by the incident and reflected wavefronts, we have

$$\sin i = \frac{A'C'}{AC'} = \frac{AC''}{AC'} = \sin i'$$

The angle between the boundary and the wavefront is equal to the angle between the normal to the boundary and the normal to the wavefront. But this latter direction represents what would be called the ray direction (at least in optics), i.e., the direction of a narrow beam of progressive waves.[2] Thus the angles i and i' represent the angles of incidence and reflection for rays encountering a straight boundary.

(Actually, the above discussion of the reflection process may seem hard to reconcile with the property, normally required of Huygens' secondary wavelets, that there should be a vanishingly small amplitude in the backward direction. One can argue, however, that the presence of a sharp boundary does create a new and different situation, in which the production of strong backward wavelets becomes possible.)

The process of refraction is analyzed in a similar way. Referring again to Fig. 8–9(b), we must specify the radius of the Huygens wavelet that has advanced from A into the second medium from the time the wavefront was at AA' to the time when it touches the boundary at C'. Then $C'C'''$, drawn tangent to this wavelet (and to all the other wavelets at this instant), is the wavefront in medium 2. If the wave velocities in the two media are

[1] You should satisfy yourself that this is so.

[2] This orthogonality of ray direction and wavefront may seem obvious, but actually ceases to hold in anisotropic media, in which the Huygens wavelets may be elliptical instead of circular.

v_1 and v_2, respectively, and if the time involved is Δt, we have

$$A'C' = v_1 \Delta t \qquad AC''' = v_2 \Delta t$$

The angle of refraction, r, is the angle between $C'A$ and $C'C'''$, and from the geometry we have

$$\sin i = \frac{v_1 \Delta t}{AC'} \qquad \sin r = \frac{v_2 \Delta t}{AC'}$$

Therefore,

$$\frac{\sin i}{\sin r} = \frac{v_1}{v_2} \tag{8-12}$$

The problem of calculating the actual amplitudes of the reflected and transmitted waves is not a trivial one. Indeed, it is not a single problem. Longitudinal (compressional) waves behave differently from transverse waves, and with transverse waves, furthermore, the case in which the displacement is perpendicular to the plane of Fig. 8–9 (as it would be with water waves) differs from that in which the displacement lies in the plane of the figure. That is, with transverse waves the behavior depends on the state of polarization. Because of these complexities, we shall not attempt to analyze such problems. But it may be noted that at normal incidence ($i = 0$) we have an essentially one-dimensional problem once again. A distinction between longitudinal and transverse disturbances may still remain, however, as we have already mentioned (p. 265) for the case of a fluid medium in contact with an effectively immovable but smooth solid boundary. There will be effectively 100% reflection of any incident wave. But if the wave is longitudinal, the boundary acts as one that is completely rigid and the reflected wave displacement at the boundary must be equal and opposite to that of the incident wave; whereas if the wave is transverse, the boundary offers no resistance and the reflection takes place without any reversal of sign of the displacement (see our earlier discussions of one-dimensional boundary problems).

In Fig. 8–10 we show examples of the reflection and refraction of water waves, as observed in a ripple tank. In the refraction, one can clearly see the change of wavelength (by the factor v_2/v_1) that occurs as the disturbance passes into the second medium.

We shall not present here any discussion of the reflection or refraction of circular waves at straight or curved boundaries. Such situations can, however, be nicely analyzed in terms of the behavior of Huygens wavelets. One sees clearly how mirrors and lenses modify incident wavefronts leading to focusing or defocus-

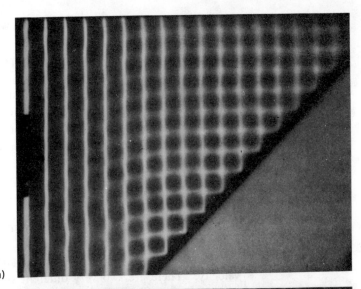

(a)

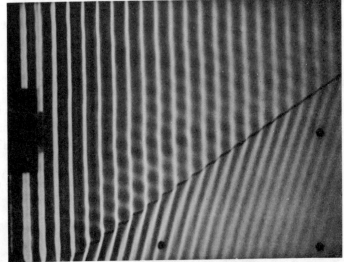

(b)

Fig. 8–10 (a) Reflection of straight water waves at a rigid boundary (i = 45°). (b) Refraction of water waves at the boundary between regions of different water depth and hence different wave speed. (From the film "Ripple Tank Phenomena," Part I, Education Development Center, Newton, Mass.)

ing effects and other such phenomena, also describable as modifications of the paths of rays according to Snell's laws.

On the specific question of refraction, it is perhaps worth pointing out that a change of direction of the wavefront occurs whenever the velocity of the waves varies with position. This can happen within a single medium under certain conditions. For example, the speed of compressional waves (sound) in gases is a function of temperature. [Specifically $v \sim \sqrt{T}$—see Chapter 7, Eq. (7–12)]. Thus, if there is a temperature gradient in a gas, waves traveling through the gas will be progressively bent. Again, if the medium itself is in motion in such a way that different parts have different velocities, refraction will occur. In air near the earth's surface there may be both a temperature gradient and a velocity gradient. Depending on their signs, and on the direction of propagation of the waves, a train of sound waves may either be bent up away from the earth's surface, or alternatively may be caused to hug the surface. In the latter case there may be an enhanced audibility of the sound over considerable distances.

DOPPLER EFFECT AND RELATED PHENOMENA

If the source of a periodic disturbance moves with respect to a medium, the pattern of waves produced by it is modified. The simplest case is of a source moving in a straight line at constant velocity. There are two very different situations, according to whether the speed of the source is less or greater than the speed of the waves that it generates. These two situations are shown schematically in Fig. 8-11. The position of the source, S, is shown at a succession of equal intervals of time. These could, for example, be instants at which the source generates a brief pulse, or instants separated by one period of a smooth sinusoidal vibration of the source. In any case, a circle with a given position of S as center represents the locus of points influenced at a given subsequent instant by waves spreading out from S.

Fig. 8-11 Successive wavefronts produced at equal time intervals from (a) source moving at less than wave speed, (b) source moving at more than wave speed.

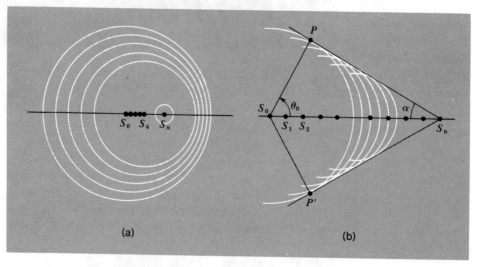

(a)   (b)

Let the speed of the source be u, and let the wave speed be v. Then at a time t after the initiation of one of the circular waves, the radius of the wavefront is vt and the source has moved a distance ut. If $u < v$, we have a situation as shown in Fig. 8-11(a). The circular wavefronts lie one inside another. The distance between successive wavefronts is least along the direction of motion of the source and greatest at $180°$ to this direction. If τ is the time interval between the successive positions of S shown in Fig. 8-11(a), then these separations of the wavefronts are $(v - u)\tau$ and $(v + u)\tau$. But $v\tau$ represents the distance between wavefronts in any direction if the source is stationary. Thus there is a systematic variation of wavelength with direction for the waves emitted from a moving source; this is the Doppler effect. In particular, we have

$$\lambda_{\min} = \lambda_0 \left(1 - \frac{u}{v}\right) \qquad \lambda_{\max} = \lambda_0 \left(1 + \frac{u}{v}\right)$$

The situation is more complicated for other directions, but can be simply analyzed if the distance from the source to the point of observation is very large compared to one wavelength. We then have a situation of the kind shown in Fig. 8-12. The points marked S_0 and S_n represent the positions of the source at $t = 0$ and at $t = n\tau$ (n periods later). Since the speed of the source is u, we have

$$S_0 S_n = x_n = un\tau$$

Since the point of observation, P, is assumed to be far away, the angle $S_0 P S_n$ is very small. This means that the wavefronts arriving at P from S_0 and S_n (and all intermediate source points) are almost parallel. Suppose that the wave W_0 from S_0 has just

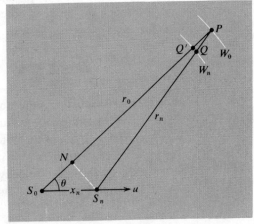

Fig. 8-12 Waves
arriving at a distant
point P from a source
moving from S_0 to S_n.

reached P. This defines a time t_P equal to r_0/v. The wave from S_n started out at $t = n\tau$; thus at time t_P it has been traveling only for a time $t_p - n\tau$; its wavefront is at W_n, and we have

$$S_n Q = v(t_P - n\tau)$$
$$= r_0 - vn\tau$$

The distance between the wavefronts can be taken to be equal to either QP or $Q'P$ (the difference between them is insignificant). If we put $S_n P = r_n$, we have

$$QP = r_n - S_n Q = r_n - r_0 + vn\tau$$

But if we drop a perpendicular from S_n onto the line $S_0 P$, we also have $NP \approx r_n$ (again because of the smallness of the angle $S_0 P S_n$), so that

$$r_0 - r_n \approx S_0 N = x_n \cos \theta$$

i.e.,

$$r_0 - r_n \approx un\tau \cos \theta$$

Substituting this in the preceding expression for QP, we have

$$Q'P \approx QP \approx vn\tau - un\tau \cos \theta$$
$$= n\lambda_0 \left(1 - \frac{u \cos \theta}{v} \right)$$

But $Q'P$ or QP spans n wavelengths of the disturbance as observed in the direction θ to the moving source. Thus we have

$$\lambda(\theta) = \lambda_0 \left(1 - \frac{u \cos \theta}{v} \right) \tag{8-13}$$

What this means, very simply, is that the Doppler effect depends on the component of source velocity in the direction of the observer. The *frequency* at which successive wavefronts pass through the point of observation P is the wave speed divided by the wavelength. Thus we have

$$\nu(\theta) = \frac{\nu_0}{1 - \dfrac{u \cos \theta}{v}} \tag{8-14}$$

This last equation is the most appropriate statement of the Doppler effect in acoustics, because the effect is detected through the change in pitch of the note received from a moving source.

Let us turn now to the case in which the source velocity exceeds the wave velocity. This gives us a situation like that shown in Fig. 8–11(b). Suppose that the source is at S_0 at $t = 0$.

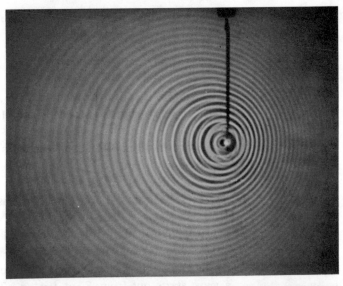

(a)

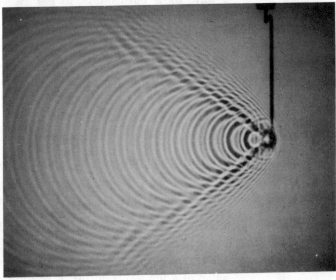

*Fig. 8–13   Water
waves produced in a
ripple tank by a mov-
ing source. (From
the film "Ripple Tank
Phenomena," Part
III, Education De-
velopment Center,
Newton, Mass.)
(a) Source speed less
than wave speed
(Doppler effect).
(b) Source speed
greater than wave
speed (shock wave).*

(b)

Then at the later time $t = n\tau$, the source is at S_n, where
$S_0S_n = nu\tau$, and the wavefront from S_0 has attained a radius
of $nv\tau$. At the position of S_n at this instant, the waves are only
just beginning to be generated. If tangent lines are drawn from
S_n to the circular wavefront from S_0, these lines are also tangent
to all the other intermediate circles. Our earlier experience with
the Huygens construction would suggest that the result is a reinforcement of the wavelets along these lines, which thus act as
straight wavefronts traveling outward at the speed v. The angle

277   Doppler effect and related phenomena

α which these wavefronts make with the line of motion of the source is defined through the relation

$$\sin \alpha = \frac{S_0 P}{S_0 S_n} = \frac{v}{u} \tag{8-15}$$

The ratio u/v is called the Mach number, and the angle α is the Mach angle (which exists only if the Mach number is greater than 1).

Figure 8–13 shows actual examples of the wave patterns generated in a ripple tank by a moving source for Mach numbers less than and greater than 1.

To see more explicitly how the locus of the circular waves for $u > v$ acts as a concentrated straight wavefront, consider the times of arrival of the successive circular waves at a point P far away from the moving source. We can refer again to Fig. 8–12. Again suppose that a wave starts out from S_0 at $t = 0$, and that a wave starts out from S_n at $t = n\tau$. The times of arrival of these waves at P are given by

$$t_0 = \frac{r_0}{v}$$

$$t_n = n\tau + \frac{r_n}{v}$$

Thus

$$t_n - t_0 = n\tau - \frac{r_0 - r_n}{v}$$

We shall again put $r_0 - r_n \approx x_n \cos \theta = nu\tau \cos \theta$, giving

$$t_n - t_0 \approx n\tau \left(1 - \frac{u \cos \theta}{v}\right)$$

Clearly if $u < v$, t_n is always greater than t_0—i.e., the waves arrive in the same order in which they are emitted. But if $u > v$, the time sequence depends on θ. And, in particular, there is a value of θ for which all the wavefronts arrive at P at the *same* instant. Calling this angle θ_0, we have

$$\cos \theta_0 = \frac{v}{u} \tag{8-16}$$

This value of θ is the complement of the Mach angle, and defines the direction, perpendicular to the straight wavefront itself, along which this region of concentration of the circular wavelets travels. In such terms we can understand the production of effects like sonic booms. If a source S [Fig. 8–14(a)] is traveling at a speed greater than the wave speed, and an observer is at P, then a line

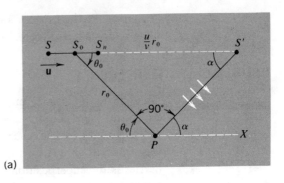

(a)

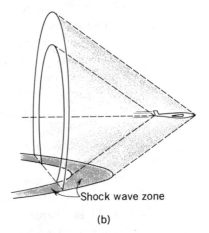

Fig. 8-14 (a) At
the direction
$\theta_0 = cos^{-1}(v/u)$,
pulses from the mov-
ing source $(u > v)$
pile up simultaneously
at the observation
point P. (b) Pro-
duction of sonic
booms.

Shock wave zone

(b)

drawn from P at an angle θ_0 to the direction of motion of the
source will intersect the line of motion of the source at a point S_0.
At a time r_0/v after the source passes through S_0, P will suddenly
receive the pile-up of wavelets which are generated by the source
over a short distance from S_0 onward, but which reach P simultaneously. [At this instant, the source itself has traveled a distance
ur_0/v beyond S_0—see Fig. 8-14(a).] Prior to this instant, P was
receiving no disturbances. After the pile-up has traveled beyond
P, there will continue to be an arrival of normal wavelets—but
without benefit of reinforcement through simultaneous arrival
they may be too weak to be noticeable.

In practice, an airplane traveling at supersonic speed generates a double boom, owing to the formation of two principal
shock fronts, one at its nose and the other at its tail. These, for
a plane traveling horizontally, at constant velocity, are in the
form of conical surfaces that are carried along with the plane
[see Fig. 8-14(b)]. Their intersection with the ground is hyperbolic

279   Doppler effect and related phenomena

in shape. As this pattern sweeps over any particular point, the sonic boom is heard there.[1]

DOUBLE-SLIT INTERFERENCE

We shall now consider more explicitly what happens when an advancing wave is obstructed by barriers. From the standpoint of Huygens' principle, each unobstructed point on the original wavefront acts as a new source, and the disturbance beyond the barrier is the superposition of all the waves spreading out from these secondary sources. Because all the secondary sources are driven, as it were, by the original wave, there is a well-defined phase relationship among them. This condition is called *coherence*, and it implies in turn a systematic phase relation among the secondary disturbances as they arrive at any more distant point. As a result there exists a characteristic interference pattern in the region on the far side of the barrier.

The simplest situation, and one that is basic to the analysis of all others, is to have the original wave completely obstructed except at two arbitrarily narrow apertures. In a two-dimensional system these then act as point sources. The analogous situation for waves in three dimensions is to have two long parallel slits which act as line sources. We briefly discussed such an arrangement in Chapter 2, when first considering the superposition of harmonic vibrations, and you are probably familiar with it also in connection with Thomas Young's historic experiment (performed about 1802) that displayed the interference of light waves in an unmistakable fashion.

In Fig. 8–15 we indicate a wavefront approaching two slits S_1 and S_2, which are assumed to be very narrow but equal. For simplicity we shall suppose that the slits are equally far from some point which acts as the primary source of the wave. Thus the secondary sources S_1 and S_2 are in phase with one another. If the original wave is a continuing simple harmonic disturbance, S_1 and S_2 in turn generate simple harmonic waves. At an arbitrary point P, the disturbance is obtained by adding together the contributions arriving at a given instant from S_1 and S_2. In general, we need to consider two characteristic effects:

1. The disturbances arriving at P from S_1 and S_2 are different in amplitude, for a dual reason. First the distances r_1 and r_2 are

[1]For a fuller account, see, for example, the article "Sonic Boom" by H. A. Wilson, Jr., *Scientific American*, Jan. 1962, pp. 36–43.

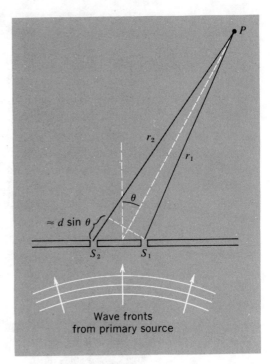

Fig. 8–15 Double-slit interference.

different, and the amplitude generated by an expanding circular disturbance falls off with increasing distance from the source. Second, the angles θ_1 and θ_2 are different, and a Huygens wavelet has an amplitude which falls away (as discussed earlier, in connection with Fig. 8–8) with increasing obliquity.

2. There is a phase difference between the disturbances at P, corresponding to the time difference $(r_2 - r_1)/v$, where v is the wave speed.

We shall concentrate on situations for which the distances r_1 and r_2 are large compared to the distance d between S_1 and S_2. Then the difference between the *amplitudes* due to S_1 and S_2 at P is negligible. But there remains the possibility of an important *phase* difference between the two disturbances, and it is this which dominates the general appearance of the resultant wave pattern. We see the typical consequences in Fig. 8–16, which is a ripple-tank photograph. There exist loci—*nodal lines*—along which the resultant disturbance is almost zero at all times. It is easy to calculate their positions. At any point such as P in Fig. 8–15, the displacement as a function of time is of the form

$$y_P(t) = A_1 \cos \omega \left(t - \frac{r_1}{v} \right) + A_2 \cos \omega \left(t - \frac{r_2}{v} \right) \qquad (8\text{–}17)$$

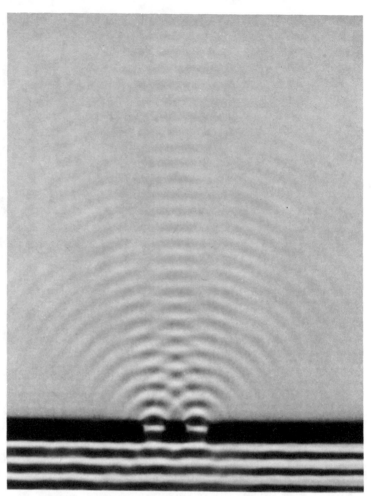

Fig. 8-16 Double-
slit interference of
water waves. (From
the film "Ripple Tank
Phenomena," Part II,
Education Develop-
ment Center, Newton,
Mass.)

if the time dependence of the disturbances at S_1 and S_2 is as
$\cos \omega t$. Equation (8–17) embodies the fact that a given sequence
of displacements at either source gives rise, at a time r/v later, to
a similar sequence at a point distance r away. Thus if we can put
$A_1 \approx A_2 (= A_0$, say), then

$$y_P(t) = A_0 \left[\cos \omega \left(t - \frac{r_1}{v} \right) + \cos \omega \left(t - \frac{r_2}{v} \right) \right]$$

$$= 2A_0 \cos \omega t \cos \left[\frac{\omega}{2v} (r_2 - r_1) \right]$$

Introducing the wavelength $\lambda = v/\nu = 2\pi v/\omega$, we thus have

$$y_P(t) = 2A_0 \cos \omega t \cos \left[\frac{\pi(r_2 - r_1)}{\lambda} \right] \tag{8–18}$$

282   Boundary effects and interference

A given nodal line is defined by the condition that the quantity $\pi(r_2 - r_1)/\lambda$ is some odd multiple of $\pi/2$. Thus we can put

$$\frac{\pi(r_2 - r_1)}{\lambda} = (2n + 1)\frac{\pi}{2}$$

or

$$r_2 - r_1 = (n + \tfrac{1}{2})\lambda \qquad \text{(nodal lines)} \qquad (8\text{-}19)$$

where n is any positive or negative integer (or zero). The nodal lines are thus a set of hyperbolas, which divide up the whole region beyond the slits in a well-defined way. Within the areas between the nodal lines, one can draw a second set of hyperbolas which define lines of *maximum* displacement—in the sense that, at a given distance from the slits, and between two given nodal lines, the amplitude of the resultant disturbance reaches its greatest value. It is easy to see that the condition for this to occur is

$$r_2 - r_1 = n\lambda \qquad \text{(interference maxima)} \qquad (8\text{-}20)$$

The important parameter that governs the general appearance of the interference pattern is the dimensionless ratio of the slit separation d to the wavelength λ. This fact is manifested in its simplest form if we consider the conditions at a large distance from the slits—i.e., $r \gg d$. Then (referring back to Fig. 8–15) the value of $r_2 - r_1$ can be set equal to $d \sin \theta$ with negligible error. Hence the condition for interference maxima becomes

$$d \sin \theta_n = n\lambda \qquad \sin \theta_n = \frac{n\lambda}{d} \qquad (8\text{-}21)$$

and the amplitude at some arbitrary direction is given by

$$A(\theta) = 2A_0 \cos\left(\frac{\pi d \sin \theta}{\lambda}\right) \qquad (8\text{-}22)$$

We see from this that the interference at a large distance from the slits is essentially a directional effect. That is, if the positions of nodes and interference maxima are observed along a line parallel to the line joining the two apertures, the *linear* separations of adjacent maxima (or zeros) increase in proportion to the distance from the slits.

The general features of the interference pattern for a double-slit system are nicely illustrated in Fig. 8–17 for two different values of d/λ. These are not real wave patterns but simulated ones, obtained by superposing two sets of concentric circles.[1]

[1]Done with items from "Moiré Patterns" kit, made by Edmund Scientific Co., Barrington, N.J.

(a)

Fig. 8–17 Moiré patterns approximating double source interference. (Done with items from "Moiré Patterns" kit, distributed by Edmund Scientific Co., Barrington, N. J.) (Photo by Jon Rosenfeld, Education Research Center, M.I.T.)

(b)

A special interest often attaches to the case when d/λ is very large. This is especially so in optical interference, where the wavelength ($\sim 6 \times 10^{-7}$ m) is likely to be extremely small compared to the slit separation (typically ~ 0.1 mm). Under these conditions ($\lambda/d \approx 10^{-2}$) we can replace $\sin \theta_n$ by θ_n itself in Eq. (8–21), so that the angular separation between any two successive interference maxima becomes just λ/d, very nearly. Furthermore, at a given distance D from the slits, the successive interference maxima are equally spaced, with a separation $D\lambda/d$.

MULTIPLE-SLIT INTERFERENCE (DIFFRACTION GRATING)

In discussing the double-slit problem we have indicated in some detail how the interference pattern is formed. But in more complicated situations we shall limit ourselves to considering the state of the interference at distances that are large compared to the

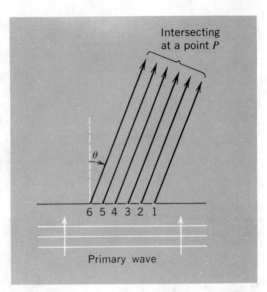

Intersecting
at a point P

θ

6 5 4 3 2 1

Primary wave

Fig. 8–18 Multiple-slit interference.

linear dimensions of the system of apertures. This permits us to assume the following:

1. Equally wide (unobstructed) portions of the original wavefront give contributions of equal amplitude at any point considered.

2. The lines to a given observation point from the various unobstructed parts of the original wavefront are almost parallel.

Let us analyze in these terms the interference pattern due to an array of N equally spaced slits. As with the double-slit problem, we shall assume for the moment that the individual slits all have the same very small width. Let the spacing between adjacent slits be d. We shall assume that the various slits are all driven in phase, as they would be if the primary wave were straight (i.e., from a very distant primary source) and parallel to the plane of the slits (Fig. 8–18). The difference in paths for secondary waves arriving at a point P from adjacent slits is equal to $d \sin \theta$. This then defines a time difference $d \sin \theta / v$ and a phase difference δ given by

$$\delta = \frac{\omega d \sin \theta}{v} = \frac{2\pi d \sin \theta}{\lambda} \tag{8–23}$$

The resultant displacement at P is thus of the form

$$y_P(t) = A_0 \cos(\omega t - \varphi_1) + A_0 \cos(\omega t - \varphi_1 - \delta)$$
$$+ A_0 \cos(\omega t - \varphi_1 - 2\delta) + \cdots \text{(to } N \text{ terms)}$$

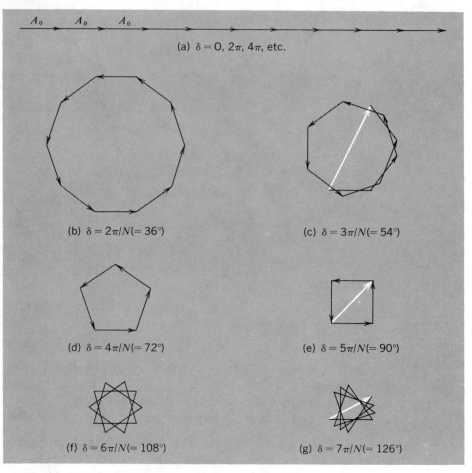

(a) $\delta = 0, 2\pi, 4\pi$, etc.

(b) $\delta = 2\pi/N (= 36°)$

(c) $\delta = 3\pi/N (= 54°)$

(d) $\delta = 4\pi/N (= 72°)$

(e) $\delta = 5\pi/N (= 90°)$

(f) $\delta = 6\pi/N (= 108°)$

(g) $\delta = 7\pi/N (= 126°)$

Fig. 8-19   Vector diagrams for diffraction grating
$(N = 10)$. (a) $\delta = 0, 2\pi, 4\pi$, etc. (b) $\delta = 2\pi/N$
$(= 36°)$. (c) $\delta = 3\pi/N (= 54°)$. (d) $\delta = 4\pi/N$
$(= 72°)$. (e) $\delta = 5\pi/N (= 90°)$. (f) $\delta = 6\pi/N$
$(= 108°)$. (g) $\delta = 7\pi/N (= 126°)$.

where $\varphi_1 = 2\pi r_1/\lambda$ is the phase difference corresponding to the distance r_1 from the first slit to the point P.

We have already considered this superposition problem in Chapter 2. The amplitude A of the resultant is obtained by taking the vector sum of N vectors of length A_0, each of which makes an angle δ with its next neighbor (see Fig. 2-7). The result is

$$A = A_0 \frac{\sin(N\delta/2)}{\sin(\delta/2)} \tag{8-24}$$

Now let us consider how A depends on the angle θ, given the equation (8–23) for δ. It is especially illuminating to do this with the help of a series of vector diagrams, such as those shown in Fig. 8–19 for the particular case $N = 10$.

1. When $\delta = 0$, the combining vectors are all in line and add together:

$$A = NA_0$$

This therefore represents the biggest possible resultant amplitude. It occurs also for every value of θ given by Eq. (8–21). That is, an array of N slits, of spacing d, has what are called *principal maxima* at the same directions as a two-slit system of the same spacing.

2. When $\delta = 2\pi/N, 4\pi/N, 6\pi/N$, etc., the combining vectors form a closed polygon and we have

$$A = 0$$

We can see this equally well from Eq. (8–24), because in all these cases the angle $N\delta/2$ is an integral multiple of π, making the numerator zero.

3. In between these zeros there will be values of δ, and hence of θ, that define intermediate maxima of displacement. These are called *subsidiary maxima* of the multiple-slit interference pattern, and their amplitudes are much less than those of the principal maxima—although their precise angular positions and relative amplitudes are not very readily evaluated, as you will discover if you try to calculate the maximum values of A from Eq. (8–24). In Fig. 8–19, the amplitude in diagram (c) for $\delta = 3\pi/N$ is *approximately* equal to that of the first subsidiary maximum, and is only about one-fifth of that of the principal maximum.

4. After $N - 1$ zeros, and $N - 2$ subsidiary maxima, we arrive at the value $\delta = 2\pi$, which defines the next principal maximum of the diffraction pattern.

Figure 8–20 is a comparison of the variations of amplitude with δ for a double-slit and a 10-slit system with equal interslit spacings. (Note the difference of vertical scales.) The "bouncing-ball" appearance of these curves is the result of taking A to be always positive, whereas Eq. (8–24) would define alternate positive and negative values between successive pairs of zeros. The effect of using more slits is to sharpen up the principal maxima. It is precisely this property, of course, that makes a diffraction grating a valuable tool in spectroscopy, because it implies a very sharp angular resolution for light of a given wavelength. Most of the intensity is concentrated within narrow angular ranges

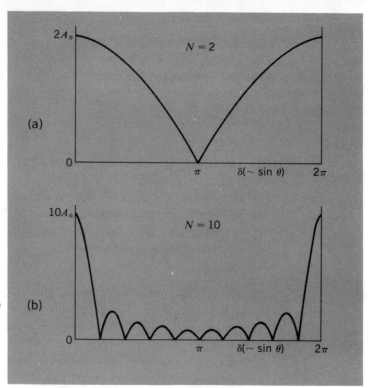

Fig. 8-20 (a) Varia-tion of amplitude with phase difference for two-slit interference. (b) Variation of amplitude with phase difference for ten-slit interference.

around the directions of the principal maxima—the so-called zero-order (straight through), first-order, second-order, etc., max-ima of the pattern. Figure 8–21 shows a multiple-slit interference pattern ($N = 8$) produced in a ripple tank. The effective con-centration of the waves into just three beams—one of zero order and two of first order—is clearly shown. (Why are no higher orders present?)

DIFFRACTION BY A SINGLE SLIT

It is clear that no individual slit or aperture can be arbitrarily narrow, and this fact gives rise to characteristic interference be-havior from the various regions of one slit alone. We have re-frained from discussing this earlier because the analysis of the N-slit problem provides some valuable background.

Figure 8–22 is a greatly enlarged diagram of an individual narrow slit, of breadth b. We assume that all parts of it are driven in phase by an incident plane wave. Now if the disturbance on the far side of the slit is to be studied at an angle θ to the normal, as shown, there is a net path difference of $b \sin \theta$ from the two

Fig. 8–21 Eight-slit interference of water waves. (From the film "Ripple Tank Phenomena," Part II, Education Development Center, Newton, Mass.)

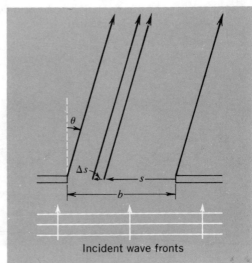

Incident wave fronts

Fig. 8–22 Single-slit diffraction.

sides of the slit to the point of observation, and an associated phase difference of $2\pi b \sin\theta/\lambda$. If we imagine the slit divided up into a large number of strips of equal width Δs, any one of these, a distance s from one extreme edge of the slit, produces at the observation point a displacement proportional to Δs with a phase (relative to waves from the edge of the slit) equal to $2\pi s \sin\theta/\lambda$. If we accepted this description of the situation, we could find the resultant amplitude as a function of θ by constructing a vector diagram just like those in Fig. 8–19 for the diffraction grating. It would correspond to putting $N = s/\Delta s$, and $\delta = 2\pi \Delta s \sin\theta/\lambda$. But, of course, this subdivision into a finite number of strips is artificial. What we must do is to imagine the limit of this description as $\Delta s \to 0$ and $N \to \infty$. We then have a continuous variation of phase in proportion to distance across the slit. The implication of this is that our vector diagram becomes a smooth circular arc, with the following properties:

1. The angle between the tangents at its two ends is the total phase difference $2\pi b \sin\theta/\lambda$.

2. The length of the arc corresponds to the total amplitude that the slit would provide (for given values of r and θ) if all parts of the slit could somehow produce their effects in phase with one another. If the obliquity factor in the Huygens wavelets is ignored, this arc length is always equal to the amplitude A_0 produced (at a given distance r from the slit) for $\theta = 0$.

The calculation of the resultant amplitude is now a straightforward matter. In Fig. 8–23 we indicate the basis of the cal-

Fig. 8–23   *Vector diagrams for single-slit diffraction.*

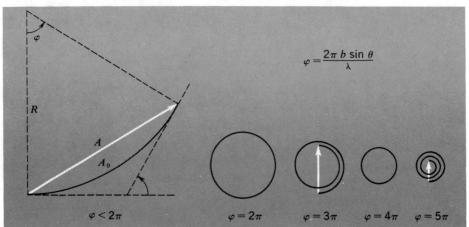

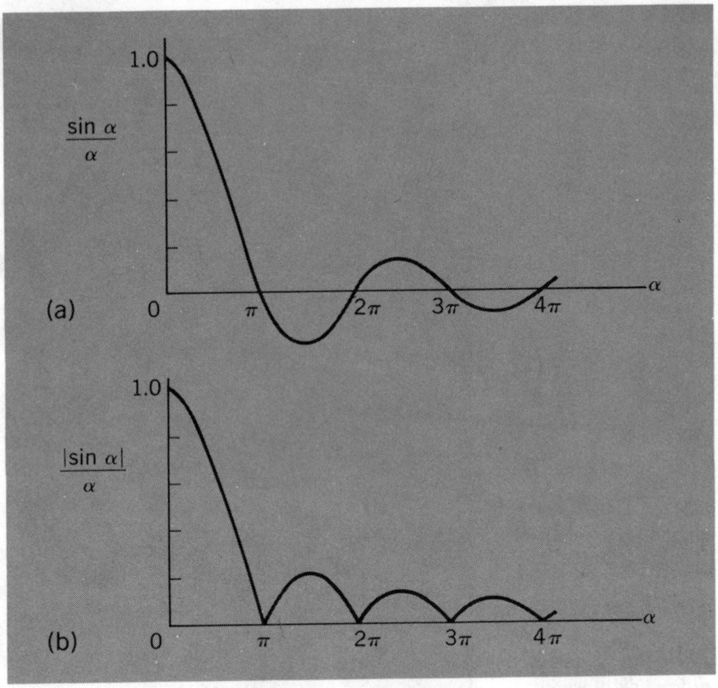

Fig. 8–24 Variation
of amplitude with di-
rection in single-slit
diffraction.
($\alpha = \pi b \sin \theta / \lambda$,
where θ is the direc-
tion of observation
and b is the slit
width.) (a) Ampli-
tude together with
phase (as shown by +
or − value).
(b) Absolute magni-
tude of amplitude.

culation. For a given value of the total phase difference φ, the
vector diagram becomes a circular arc of radius R such that

$$A_0 = R\varphi$$

The resultant amplitude A under these conditions is the chord
of this arc and hence is given by

$$A = 2R \sin(\varphi/2)$$

Thus we have

$$A = A_0 \frac{\sin(\varphi/2)}{\varphi/2} \qquad \text{where } \frac{\varphi}{2} = \frac{\pi b \sin \theta}{\lambda} \qquad (8\text{–}25)$$

This variation of resultant amplitude with direction is thus of the
form $(\sin \alpha)/\alpha$, where $\alpha = \varphi/2$. This function (more formally
identified as a Bessel function of order zero) has a zero whenever
$\varphi/2$ is an integral multiple of π. Its general appearance is shown
in Fig. 8–24(a). In Fig. 8–24(b) it is replotted without regard to
sign, and its close resemblance to the amplitude curve for a
diffraction grating [Fig. 8–20(b)] is then more readily appreciated.

It follows from this analysis that one slit, alone, can give rise
to a diffraction pattern with a system of nodal lines, as shown in
Fig. 8–25. It is essentially like the pattern around the central
(zero-order) maximum of a diffraction grating, rather than a

291   Diffraction by a single slit

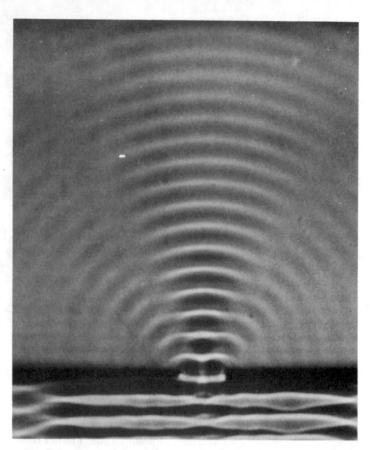

Fig. 8–25 Single-slit diffraction of water waves. (From the film "Ripple Tank Phenomena," Part II, Education Development Center, Newton, Mass.)

double-slit pattern. The subsidiary maxima are relatively feeble; their amplitudes are approximately proportional to the values of $(\sin \alpha)/\alpha$ for $\alpha = 3\pi/2$, $5\pi/2$, etc., i.e., for $\varphi = 3\pi$, 5π, etc. (Not exactly, because the maxima do not occur at precisely these phase values.) Relative to an amplitude 1 for $x = 0$, these other maxima would thus have amplitudes of about $2/3\pi$ (0.21), $2/5\pi$ (0.13), etc. (see Fig. 8–24).

Note that the first zeros occur for directions such that the path difference from the two sides of the slit is just one complete wavelength. This makes good sense if one imagines the single slit as being made up of two contiguous slits, each of width $b/2$. The path difference between waves from the centers (or other pairs of corresponding points) of these two parts is then $\lambda/2$, which is the condition for destructive interference. The other nodal lines for a single slit can be understood in a similar way.

It should always be remembered that all our discussion pertains to points of observation that are far from the slit or slits.

This is particularly important now that we have recognized the consequences of finite slit width. For, of course, at positions close to a slit, we see the effect of slit width in a much more direct way. A portion, of width b, of the incident wavefront is permitted to pass through, and a strong disturbance exists over this region, whereas all other points on the far side of the barrier are in the geometrical shadow. How far away must we go before our description in terms of angles of diffraction takes over? We can establish a criterion, as follows: The central maximum of the diffraction pattern of a slit of width b extends over a range of angles $\pm\theta_m$, where

$$\sin \theta_m = \frac{\lambda}{b}$$

[This is implied by Eq. (8–25).] At a distance D from the slit (Fig. 8–26), this maximum would define a linear spread equal to $\pm D \tan \theta_m$. On the other hand, a purely geometrical image of the slit would always be of width b. Thus the diffraction is dominant if the following condition holds:

$$D \tan \theta_m \gg b$$

If λ is small compared to b, we can put $\tan \theta_m \approx \sin \theta_m = \lambda/b$, and our condition becomes

$$D \gg \frac{b^2}{\lambda} \tag{8–26}$$

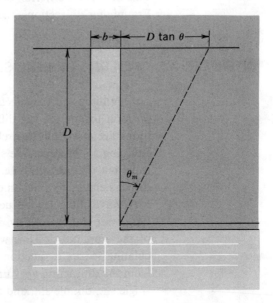

Fig. 8–26 Conditions for Fraunhofer diffraction.

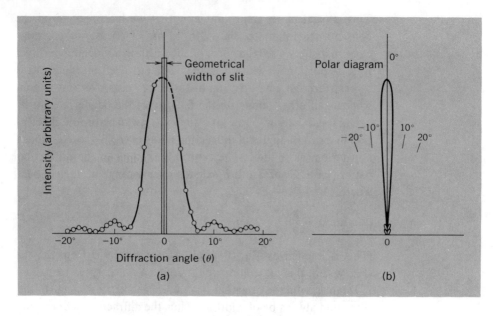

Fig. 8–27 (a) Single-slit diffraction of sound waves (λ = 1.45 cm). (b) Polar diagram of same pattern. Length of line from O to the curve in any direction gives relative intensity in that direction. (After R. W. Pohl, Physical Principles of Mechanics and Acoustics, *Blackie, London, 1932.)*

This important criterion defines the conditions for what is called Fraunhofer diffraction—the type we have been discussing.

INTERFERENCE PATTERNS OF REAL SLIT SYSTEMS

Having discussed the effects of finite slit width, we are now in a position to analyze the diffraction or interference patterns (the terms are essentially interchangeable) of any perforated barrier. In doing so, however, we shall consider not the resultant amplitude, but the *intensity*—i.e., the rate at which the resultant wave delivers energy to a region of a given size at various points. Now for a wave of a given frequency or wavelength in a given medium, the power transported by the wave is proportional to the square of the amplitude. Thus we are essentially concerned with calculating A^2 as a function of direction at some given distance from the diffracting aperture. We shall take the specific cases of single slit, double slit, and multiple slit (grating).

1. *Single slit.* For a single slit, on the basis of Eq. (8–25), we have

$$I(\theta) = I_0 \left(\frac{\sin \alpha}{\alpha}\right)^2 \quad \text{where } \alpha = \frac{\pi b \sin \theta}{\lambda} \tag{8–27}$$

Figure 8–27 shows a beautiful example of such a pattern, obtained by R. W. Pohl with sound waves. The wavelength λ was 1.45 cm (corresponding to a supersonic frequency of about 23 kHz), and the slit width b was 11.5 cm. The second version of the pattern is a *polar diagram*; in this the distance measured from the origin to any point on the curve is proportional to the intensity in that particular direction.

Once one has recognized that it is A^2, rather than A itself, that provides a measure of the most important quantity—the energy flow—one appreciates better how very important the central maximum is compared to the others. The heights (theoretically) of the most important subsidiary maxima, i.e., those nearest to the central maximum, are only about 5% of the central one, and about 93% of the total transmitted energy lies between the zeros on either side of the central maximum. Incidentally, the squaring of the ordinates in curves like Fig. 8–24(b) gets rid of the discontinuities of slope at the zeros (satisfy yourself that this follows from the equations).

2. *Double slit.* In this case we have a combination of two effects—the characteristic diffraction pattern of one slit alone, and the interference between the two slits. The intensity is given by an expression of the form

$$I(\theta) = 4I_0 \left(\frac{\sin \alpha}{\alpha}\right)^2 \cos^2 \left(\frac{\delta}{2}\right) \tag{8–28}$$

where $\alpha = (\pi b \sin \theta)/\lambda$ and $\delta = (2\pi d \sin \theta)/\lambda$. Here I_0 is the maximum intensity (for $\theta = 0$) that would be obtained from one slit alone. The above equation is based on Eq. (8–22) for two slits of negligible width, combined with Eq. (8–27).

Careful measurements on a double-slit interference pattern will reveal this modulation of the basic interference effect by the single-slit pattern. The slit separation d (measured between centers) is necessarily larger (and perhaps much larger) than the width b of an individual slit, so the angular width of the single-slit modulation is significantly larger than the angular separation between the interference peaks. If the slits are extremely narrow compared to their separation, the whole double-slit pattern may

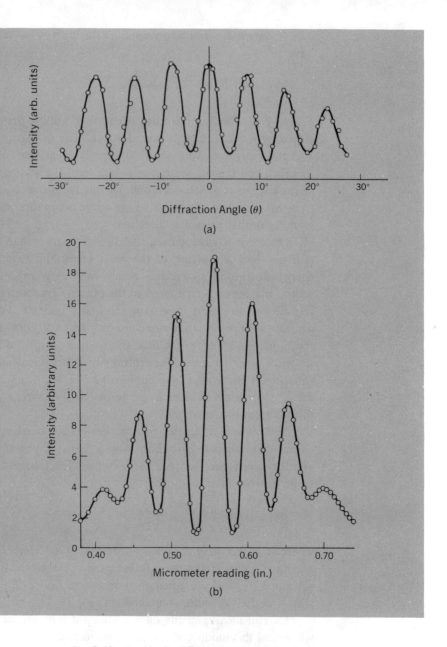

Fig. 8-28 Double-slit diffraction patterns, showing influence of width of individual slit. (a) Acoustic pattern, with λ = 1.45 cm. (After R. W. Pohl, Physical Principles of Mechanics and Acoustics, Blackie, London, 1932.) (b) Optical pattern, with λ = 7300 Å. [From A. P. French, J. G. King, and D. J. Cronin, "An Interference Fringe Photometer," Am. J. Phys., 33, 628 (1965).]

lie within the central maximum of the single-slit diffraction pattern. Figure 8–28(a) (again from Pohl) is an acoustic interference pattern obtained under such conditions ($b/d \approx 1/10$). Figure 8–28(b) is an optical double-slit pattern for which $b/d \approx \frac{1}{4}$. The limits of intensity imposed by the single-slit diffraction factor are nicely indicated in this case.

It is worth commenting on the factor 4 in Eq. (8–28). For $\theta = 0$ the intensity due to two slits is four times as great as that due to one slit by itself. Clearly, however, the total amount of energy transported by waves through two slits is only twice that passed by one slit alone. The augmentation by a factor of more than 2 in some directions is offset by the existence of zero intensity in other directions—along the nodal lines. The interference is essentially a redistribution of the available energy.

3. *Diffraction grating*. The appropriate formula in this case is the combination of Eq. (8–27) with Eq. (8–24). This gives us

$$I(\theta) = I_0 \left(\frac{\sin \alpha}{\alpha}\right)^2 \left[\frac{\sin(N\,\delta/2)}{\sin(\delta/2)}\right]^2 \tag{8–29}$$

[You can check that for $N = 2$ this reproduces Eq. (8–28).] A quantitative study of the fine details of such a multiple-slit pattern for mechanical waves (e.g., sound) is not easy; any ordinary

Fig. 8–29   Intensity pattern of acoustic diffraction grating ($N = 7$) for $\lambda = 1.45$ cm. (After R. W. Pohl, Physical Principles of Mechanics and Acoustics, Blackie, London, 1932.)

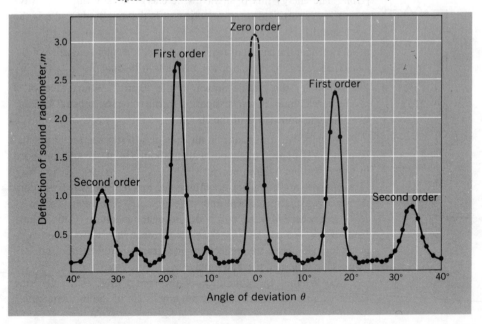

laboratory contains all kinds of extraneous surfaces and objects that scatter the waves and give background. Figure 8–29, however, shows how the main features of the expected pattern are displayed. This was obtained (by R. W. Pohl) using a grating of seven slits and sound of wavelength 1.45 cm.

As with the double-slit system, we may note the redistribution of the available energy. If we ignore the variation of the factor $(\sin \alpha/\alpha)^2$, each principal maximum reaches an intensity equal to N^2 times that due to a single slit. The width of this maximum is, however, only about $1/N$ of the separation between maxima. The combination of these factors gives an integrated intensity equal to N times that due to one slit alone.

PROBLEMS

8–1 Two strings, of tension T and mass densities μ_1 and μ_2, are connected together. Consider a traveling wave incident on the boundary. Find the ratio of the reflected amplitude to the incident amplitude, and the ratio of the transmitted amplitude to the incident amplitude, for the cases $\mu_2/\mu_1 = 0, 0.25, 1, 4, \infty$.

8–2 Two strings, of tension T and mass densities μ_1 and μ_2, are connected together. Consider a traveling wave incident on the boundary. Show that the energy flux of the reflected wave plus the energy flux of the transmitted wave equals the energy flux of the incident wave. [*Hint:* The energy flux of a wave (the energy density times the wave speed) is proportional to A^2/v, where A is the amplitude and v is the wave speed.]

8–3 Consider the circuit drawn in the figure. Calculate the value of the resistance X for maximum power dissipation through it.

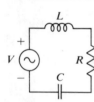

8–4 Consider the circuit drawn in the figure. What value of ω produces maximum power dissipation through the resistance R? (*Hint:* Consider the impedance of the circuit.)

8–5 A plane wave of sound in air falls on a water surface at normal incidence. The speed of sound in air is about 334 m/sec and the speed in water is about 1480 m/sec.
[The appropriate boundary conditions for longitudinal waves are continuity of wave displacement and wave pressure. The latter is given by $K(\partial \xi/\partial x)$, where K is the bulk modulus of the medium. (This follows from $\Delta p = -K\Delta V/V = -K\Delta\xi/\Delta x$.) Since the wave speed v is given by $(K/\rho)^{1/2}$, the reflection and transmission coefficients are expressible in terms of ρ and v only.]

(a) What is the amplitude of the sound wave that enters the water, expressed as a fraction of the amplitude of the incident wave?

(b) What fraction of the incident energy flux enters the water?

8-6 (a) You may have observed that water waves advancing toward shore have their wavefronts almost always parallel to the shoreline, independent of the direction of the wind. Noting the fact that the velocity of waves in water decreases as the depth of the water decreases, use Huygens' Principle to explain this phenomenon.

(b) To make the analysis of (a) more specific, assume that waves, initially traveling in the x direction, enter a region in which their speed v has a systematic variation with the distance y perpendicular to the direction of travel. (For example, x could be the direction parallel to the shore, and y would then be the direction perpendicular to the shore.) Show that the direction of the waves will begin to follow the arc of a circle of radius R, such that

$$R = \frac{v}{dv/dy}$$

8-7 (a) As was developed in the text [Eq. (7–12) p. 212], the velocity of sound in a gas is proportional to the square root of the absolute temperature T. Use this fact, and the result of the previous problem, to show that when a thermal gradient exists in the vertical direction (z) sound waves will be turned initially with a radius of curvature

$$R = \frac{2T}{dT/dz}$$

(b) On a still day, the temperature of the atmosphere is found to decrease more or less linearly with height. Sketch the paths of "rays" of sound emitted from a source suspended high in the atmosphere. Assuming that the velocity of sound at ground level is 1100 ft/sec, estimate the horizontal distance at which an airplane flying at 15,000 ft first becomes audible to an observer on the ground, if the temperature decreases by 1° C per 500-ft increase in altitude.

8-8 (a) A police car, traveling at 60 mi/hr, passes an innocent bystander while sounding its siren, which has a frequency of 2000 Hz. What is the over-all change of frequency of the siren as heard by the bystander?

(b) The police car continues down the street, the far end of which is blocked by a high brick wall. What does the bystander hear when the acoustic reflections from the wall are superposed on the sound coming directly from the siren?

8-9 Sodium atoms, thermally excited, are found to emit light of characteristic wavelength $\lambda \approx 6000\text{Å}$. The radiation from a sodium-vapor source is found not to be perfectly monochromatic, but contains a distribution of wavelengths in the range $(6000 \pm .02)\text{Å}$. If this broadening of the sodium line is due predominantly to the Doppler effect (which it is), determine the approximate temperature of the sodium source. (Speed of light $= 3 \times 10^8$ m/sec).

8–10 Lord Rayleigh, in his famous treatise *The Theory of Sound,* (Vol. II, Sec. 298) noted that an observer, if he were to travel away from a musical performance at exactly twice the velocity of sound, "would hear a musical piece in correct time and tune, but backwards." Though this certainly seems plausible, think out in detail what is involved in this amusing result.

8–11 Sound waves travel horizontally from a source to a receiver. Assume that the source has a speed u, that the receiver has a speed v (in the same direction) and that a wind of speed w is blowing from the source toward the receiver. Show that, if the source emits sound of frequency ν_0, and if the speed of sound in still air is V, the frequency recorded by the receiver is given by

$$\nu = \nu_0 \frac{V - v + w}{V - u + w}$$

Note that if the velocities of source and receiver are equal, the existence of the wind makes no difference to the observed frequency of the received signal.

8–12 The text (pp. 275–276) develops the theory of the Doppler effect for a moving source, with a distant observer at a direction θ to the motion of the source. It is shown [Eq. (8–14)] that the received frequency is given by

$$\nu(\theta) = \frac{\nu_0}{1 - \dfrac{u \cos \theta}{v}}$$

(a) Show that, if the source is at rest, and the *observer* has the velocity $-u$, so that the relative velocity of source and observer is the same as before, the frequency detected by the observer is given by

$$\nu'(\theta) = \nu_0 \left(1 + \frac{u \cos \theta}{v} \right)$$

(b) Find the approximate difference between ν and ν'. It is a matter of great importance in physics that for light waves in vacuum, in contrast to sound waves in air, there is no such difference; only the *relative* velocity of source and observer appears in the result. This is one of the features built into Einstein's special theory of relativity, according to which there is no identifiable medium with respect to which the velocity of light has some characteristic velocity.

8–13 A source of sound of frequency ν_0 moves horizontally at constant speed u in the x direction at a distance h above the ground. An observer is situated on the ground at the point $x = 0$; the source passes over this point at $t = 0$.

(a) Show that the signal received at any time t_R at the ground was emitted by the source at an earlier time t_S, such that

$$\left(1 - \frac{u^2}{v^2}\right) t_S = t_R - \frac{1}{v}\left[h^2\left(1 - \frac{u^2}{v^2}\right) + u^2 t_R{}^2\right]^{1/2}$$

(b) Show that the frequency of the received signal, as a function of the *emission* time t_S, is given by

$$\nu(t_S) = \frac{\nu_0}{1 + \frac{u}{v} \cdot \frac{u t_S}{(h^2 + u^2 t_S{}^2)^{1/2}}}$$

(The expression for ν as a function of the *reception* time t_R is considerably more complicated.)

(c) The frequency of received sound from such a source is observed to be 5500 Hz when the source is far away and approaching; it falls to 4500 Hz when the source is far away and receding. Furthermore, the frequency is observed to fall from 5100 Hz to 4900 Hz during a time of 4 sec as the source passes overhead. Deduce the speed and the altitude of the source. Approximate freely to simplify the algebra. (This kind of analysis is used to infer speed and altitude for earth satellites from the variation with time of the received frequency of a radio transmitter in the satellite.)

8–14 (a) A source, S, of sound of wavelength λ is placed a small distance d away from a flat, reflecting wall. Show that this gives rise to an interference pattern of just the kind that would be caused if the wall were absent and a second source, S', were placed a distance d behind the wall. Prove that this "image source" would have to be 180° out of phase with S, and consider what implications this has for the resulting interference pattern, as compared with that due to a normal double-source arrangement in which the two sources are in phase.

(b) If a hi-fi speaker is placed 1 ft from a wall, what range of audio frequencies will produce two or more interference fringes in a room of moderate size (e.g., 12 ft × 18 ft)? If you were sitting 12 ft from the speaker, with your head 3 ft from the wall, what frequencies would tend to be suppressed by the interference effects?

8–15 Consider an N-slit diffraction grating with slit spacing 0.05 mm and $\lambda = 5000$ Å.

(a) Approximately how many orders of principal maxima are there?

(b) What is the ratio of the two amplitudes A and A_0? (A_0 is the amplitude which would result if $N = 1$.)

(c) Show that your answer to part (b) reduces to the result derived in the text for a two-slit system if $N = 2$.

(d) If $N = 100$ find (approximately) the ratio of the amplitude of the first subsidiary maximum to that of the principal maximum.

8–16 A Fraunhofer diffraction experiment is performed using light of wavelength 5000 Å with a slit of width 0.05 mm.

(a) How far away must the detecting screen be?

(b) If a two-slit system is used, what is the ratio of intensities of the first side-maximum to the central maximum if the distance between the centers of the (identical) slits is 0.1 mm? 0.05 mm?

8–17 Sound of frequency 2000 Hz falls at normal incidence on a high wall in which there is a vertical gap, 18 in. wide. A man is walking parallel to the wall at a distance of 50 ft from it on the far side. Over what range of distance would he hear an intensity of sound more than 50% of the maximum value? More than 5%?

A short bibliography

An introduction to mechanical vibrations and waves may, of course, be found in many textbooks of general physics. Some of the older books, in particular, have good and interesting discussions of sound waves and music (for example, Lloyd W. Taylor, *Physics, The Pioneer Science*, Dover, New York, 1959, or R. A. Millikan, Duane Roller, and E. C. Watson, *Mechanics, Molecular Physics, Heat, and Sound*, M.I.T. Press, Cambridge, Mass., 1965, both reprints of books first published considerably earlier). The well-known affinity between scientists and music is apparent in these and many other sources.

The following annotated list comprises books that relate either to individual topics or to the whole scope of the present text. In general these references are comparable in level to the present book, although some of them are definitely more advanced and many of them treat individual topics in far greater detail.

Backus, John, *The Acoustical Foundations of Music*, Norton, New York, 1969.
> A book about the physics of musical sound, based on a university course for musicians, not scientists.

Barker, J. R., *Mechanical and Electrical Vibrations*, Methuen Monograph, Methuen, London, 1964 (also Wiley, New York).
> A rather thorough analytical discussion of oscillatory systems, written from a theoretical engineering standpoint.

Benade, Arthur, *Horns, Strings and Harmony*, Doubleday, Science Study Series, New York, 1960.
> A loving account, in delightfully simple terms, written by a physicist who is also a dedicated musician. Almost no mathematics, but rich in physical ideas and results.

Bishop, R. E. D., *Vibration*, Cambridge University Press, New York, 1965.

A very fine general account of vibrations, with special reference to engineering problems. Based on the 133rd set of the renowned Christmas Lectures at the Royal Institution, London.

Bland, D. R., *Vibrating Strings*, Library of Mathematics Series, Routledge and Kegan Paul, London, 1960.

A detailed mathematical analysis of vibrations and waves on strings, including some consideration of resistive and dissipative effects.

Booker, H. G., *A Vector Approach to Oscillations*, Academic, New York, 1965.

A book about the complex vector method for the analysis of oscillatory motion, showing to full advantage the power and the scope of this approach.

Braddick, H. J. J., *Vibrations, Waves, and Diffraction*, McGraw-Hill, New York, 1965.

An account that moves quickly through much the same set of theoretical topics as in the present book, but goes further, especially in the discussion of Fourier analysis and the mathematical basis of Huygens's principle.

Brillouin, L., *Wave Propagation in Periodic Structures*, Dover, New York, 1953.

A classic work on the theory of vibrations and waves in lattices, analyzed from the standpoint of circuit theory and electrical engineering but with applications to basic problems in the atomic theory of solids.

Coulson, C. A., *Waves*, Oliver & Boyd, Edinburgh, 1941 [also Wiley (Interscience), New York].

A general introduction to the mathematical theory of various kinds of waves and normal-mode problems.

Crawford, F. S., *Waves* (Berkeley Physics Series, Vol. 3), McGraw-Hill, New York, 1968.

A very thorough and rich discussion of the physics of waves, beginning with the normal-mode problem. It concerns itself extensively with electromagnetic waves and optics, as well as with mechanical waves. It is packed with sophisticated things but also with ingenious suggestions for many delightful home-and-kitchen experiments. A real tour de force.

Den Hartog, J. P., *Mechanical Vibrations*, McGraw-Hill, New York, 1956.

A well-known and excellent textbook about vibrational problems from an engineering standpoint.

Feather, N., *Vibrations and Waves*, Edinburgh University Press, Edinburgh, 1961 [also Penguin Books, London (1964)].

An extended essay, rather than a textbook, with many interesting pieces of incidental fact and comment. There are quite detailed discussions of mechanical vibrations, sound and water waves, and the phenomena of interference and diffraction.

Jeans, J. H., *Science and Music*, Cambridge University Press, New York, 1961.

A book intended primarily for the nonscientist. Almost no mathematics but much detail about the production and hearing of musical sounds.

Josephs, J. J., *The Physics of Musical Sound*, Momentum Books, Van Nostrand Reinhold, New York, 1967.

Quite similar in scope to the books by Benade and Jeans but at a higher technical and theoretical level as far as the physics is concerned.

Kinsler, L. E., and Frey, A. R., *Fundamentals of Acoustics*, Wiley, New York, 1962.

A book that closely links theory and practice in the production, transmission, and reception of sound. Aimed primarily at acoustic engineers.

Lindsay, R. B., *Mechanical Radiation*, McGraw-Hill, New York, 1960.

A detailed theoretical treatise on mechanical waves and acoustics. Quite advanced, and rich in details.

Magnus, K., *Vibrations*, Blackie, London, 1965.

A book about the mathematical analysis of mechanical vibration, with considerable attention to nonlinear systems.

McLachlan, N. W., *Theory of Vibrations*, Dover, New York, 1951.

A concise theoretical introduction to the analysis of linear and nonlinear mechanical systems.

Morse, P. M., *Vibration and Sound*, McGraw-Hill, New York, 1948.

An authoritative theoretical account of vibrating systems and the transmission and scattering of sound. Well above the level of the present book.

—— and Ingard, K. U., *Theoretical Acoustics*, McGraw-Hill, New York, 1968.

This book is basically a much expanded modern revision of the preceding reference.

Pain, H. J., *The Physics of Vibrations and Waves*, Wiley, New York, 1968.

In its general coverage this quite resembles the present text. It is more purely theoretical (and somewhat more advanced in this respect) and contains some explicit discussion of electromagnetic wave theory.

Pearson, J. M., *A Theory of Waves*, Allyn & Bacon, Boston, 1966.

A fairly sophisticated introduction to the formal theory of mechanical and electromagnetic waves.

Pohl, R. W., *Physical Principles of Mechanics and Acoustics*, Blackie, London, 1932.

A book that ties the development of the subject to observation, experiment, and demonstration in every possible way. There is very little mathematics, but the book should not be called elementary, for it is packed with physics. Based upon its author's renowned lectures at the University of Göttingen.

Rayleigh, Lord (J. W. Strutt), *The Theory of Sound*, Dover, New York, 1945.

The great classic theoretical treatise on this subject. Vol. I is concerned with vibrating systems, Vol. II with waves in fluids. The mathematical level is high, but the book is full of fascinating observational details.

Stephens, R. W. B., and Bate, A. E., *Wave Motion and Sound*, Edward Arnold & Co., London, 1950.

An interesting and extremely well organized textbook for a self-contained course on mechanical vibrations and acoustics. Somewhat above the level of the present text. It links the subject very effectively to practical applications.

Stoker, J. J., *Nonlinear Vibrations*, Wiley (Interscience), New York, 1950.

This book begins where the present book leaves off. It is concerned exclusively with the mathematical analysis of vibrating systems. For the ambitious reader only.

———, *Water Waves*, Wiley (Interscience), New York, 1957.

A very detailed and quite advanced theoretical study of water waves of all kinds.

Sutton, O. G., *Mathematics in Action*, Harper Torchbooks, New York, 1960.

An informal and delightful introduction to the use of mathematics in physical problems. It is listed here because it contains a very nice chapter entitled "An Essay on Waves."

Temple, G., and Bickley, W. G., *Rayleigh's Principle*, Dover, New York, 1956.

An introduction to the detailed mathematical analysis by which the characteristic frequencies of complicated mechanical systems can be obtained from a calculation of the total energy. (Rayleigh's principle itself states that the lowest vibrational mode of an elastic system has that distribution of kinetic and potential energies which makes the frequency a minimum.)

Timoshenko, S., *Vibration Problems in Engineering*, Van Nostrand Reinhold, New York, 1937.

A well-known older treatise on the detailed application of mathematical principles to mechanical vibrating systems.

Towne, D. H., *Wave Phenomena*, Addison-Wesley, Reading, Mass., 1967.

A detailed discussion of wave propagation, with a strong emphasis on electromagnetic waves and optics. There is a good mix of theory and experiment. Significantly above the level of the present text.

Waldron, R. A., *Waves and Oscillations*, Momentum Books, Van Nostrand Reinhold, New York, 1964.

A good brief survey of mechanical and electromagnetic waves in theory and experiment. Includes a discussion of guided waves.

Wood, A., *Acoustics*, Blackie, London, 1940.

A very thorough general account of theory and observation in acoustic vibrations and waves. It is a scholarly book in the best sense, replete with details accumulated by the author during a long and dedicated study of the subject.

—— (rev. by J. M. Bowsher), *The Physics of Music*, Methuen, London, 1961.

A book very similar to that of Jeans, but with a stronger emphasis on details and technicalities. All such books acknowledge their indebtedness to the great nineteenth-century treatise, *The Sensations of Tone*, by H. von Helmholtz.

Answers to problems

CHAPTER 1

1–4 (b) $r_1 = \sqrt{7}$, $\tan \theta_1 = \sqrt{3}/2$; $r_2 = 7$,
$\theta_2 = -2\theta_1$ ($\tan \theta_2 = -4\sqrt{3}$).
1–9 Yes; it is worth almost 21 cents.
1–10 $C = (A^2 + B^2)^{1/2}$; $\tan \alpha = -B/A$.
1–11 (a) $A = 5$ cm; $\omega = 2\pi \sec^{-1}$; $\alpha = \pm\pi/2$. (b) (For $\alpha = +\pi/2$) $x = 5\sqrt{3}/2$ cm; $dx/dt = 5\pi$ cm/sec; $d^2x/dt^2 = -10\sqrt{3}\pi^2$ cm/sec^2.
1–12 (a) $A = 150/\pi$ cm; $\omega = \pi/3 \sec^{-1}$; $\alpha = \pi/6$.
(b) $x = -75\sqrt{3}/\pi$ cm; $dx/dt = -25$ cm/sec; $d^2x/dt^2 = 25\pi/\sqrt{3}$ cm/sec^2.

CHAPTER 2

2–1 Values of (A, α) are (a) $\sqrt{2}$, $-\pi/4$; (b) 1, $-2\pi/3$; (c) $\sqrt{13}$, $-\tan^{-1}(\frac{2}{3})$; (d) $2 - \sqrt{2}$, $3\pi/4$.
2–2 $A \simeq 0.52$ mm; $\delta \simeq 33.5°$.
2–3 1 sec.
2–4 (a) $\nu = 1 \sec^{-1}$; (b) 6.25 $\sec^{-1}$; (c) 0.49 $\sec^{-1}$.

CHAPTER 3

3–1 $k = 25$ dyn/cm.
3–2 (a) $T_0 = 2\pi(m/k)^{1/2}$; (b) $T_0/\sqrt{2}$; (c) $\sqrt{2}\,T_0$.
3–3 (a) $y = 2.5$ cm; (b) 1.25 cm.

3–4 (a) $\omega = (g/l)^{1/2}$.

3–5 $2\pi(2L/3g)^{1/2}$.

3–6 $2\pi(d/g)^{1/2}$.

3–7 $y = l_0/20$; tension $= 5 \times$ weight of object.

3–8 (a) 0.25 mm; (b) 0.23 m.

3–9 (a) 22 cm radius, 360 kg; (b) 66 sec.

3–10 (a) 5.9×10^{11} N/m^2; (b) b/a; (c) 1.5.

3–11 (a) $\omega = (\gamma pA/ml)^{1/2}$.

3–14 (b) 4 N-sec/m; (b) $Q = 1$.

3–15 (a) $Q_0 = 512\pi/\log_e 2$; (b) $2Q_0$; (c) $Q = 12$,
$b = 0.025$ kg/sec.

3–16 (a) $8\pi^4\nu^3 A^2 Ke^2/c^3$; (b) $mc^3/4\pi\nu Ke^2$; (c) $(Q \log_e 2)/2\pi$;
(d) $Q \simeq 2.5 \times 10^7$; half-life $\simeq 5 \times 10^{-9}$ sec.

3–17 (d) $2\pi(2h/g)^{1/2}$.

3–19 (c) $T_x/T_y = (1 - l_0/l)^{1/2}$; (d) $x(t) = A_0 \cos(2k/m)^{1/2}t$,
$y(t) = A_0 \cos[2k(l - l_0)/ml]^{1/2}t$.

CHAPTER 4

4–3 (a) $T = \pi/5\sqrt{3}$ sec; (b) 1.3 cm.

4–4 (b) $(35g/36h)^{1/2}$; (c) $3(h/g)^{1/2}$; (d) $Q = 3$; (e) $\delta = \pi/2$;
(f) $0.90h$.

4–5 (b) 15.7 cm; (c) $\omega_0 \pm 0.017$ sec^{-1}.

4–6 (d) About 200 Å.

4–8 (b) $A = F_0/m\omega(\omega^2 + \gamma^2)^{1/2}$; $\tan \delta = -\gamma/\omega$.

4–9 (a) $\pi b\omega A^2$.

4–11 (a) 1.3 cm, 130°; (b) 0.063 J; (c) 0.30 W.

4–12 (a) 19.8 sec^{-1}; (b) 1.5 cm; (c) 0.086 W.

4–13 (a) $\omega_0 = 40$ sec^{-1}, $Q = 20$; (b) 16.

4–14 (a) $Q = 25$; (b) $\gamma = 0.04\omega_0$; (c) 0.08π; (d) $\sqrt{2}\omega_0$;
(e) $\sqrt{2}\,Q$; (f) $\bar{P}_m$; (g) E_0.

4–15 (a) $1.005\omega_1$; (b) $Q = 5$ (very nearly); (c) $0.2(mk)^{1/2}$ (approx.).

4–16 (a) $\omega_0 = (LC)^{-1/2}$; (b) $\gamma = 1/CR$; (c) $P_m = I_0^2 R/2$.

4–17 (a) $2\pi \times 10^{-5}$ J; (b) 10^{-3} J; (c) 10^{-4} sec.

CHAPTER 5

5–2 (a) 1.27 sec, 1.23 sec; (b) 40 sec (approx.).

5–4 $m\omega^2 = \left(\dfrac{k_A + k_B}{2} + k_C\right) \pm \left[\left(\dfrac{k_A - k_B}{2}\right)^2 + k_C^2\right]^{1/2}$.

If $k_C^2 = k_A k_B$, $\omega' = [(k_A + k_B + k_C)/m]^{1/2}$, $\omega'' = (k_C/m)^{1/2}$.

5–5 $\omega = \omega_0(1 \pm \alpha)^{-1/2}$.

5–6 (a) $\sqrt{6}$ sec, $3\sqrt{2}$ sec; (c) $3\sqrt{2}(\sqrt{3} + 1)/2$ sec.

5–7 (c) 2.29 sec^{-1}; (d) $k_e/k_0 = 1.52$.

5–8 (d) $(g/L)^{1/2}$; $[(g/L) + (2ka^2/mL^2)]^{1/2}$.

5–9 (d) $(\frac{11}{3})^{1/2} = 1.91$.

5–10 In "slow" mode, amplitude ratio (upper/lower) = $(\sqrt{5} - 1)/2$; in fast mode, ratio = $(\sqrt{5} + 1)/2$.

5–11 (b) $\omega^2 = [(k/2M) + (g/l)] \pm [(k/2M)^2 + (g/l)^2]^{1/2}$.

5–13 (a) Period = $2\pi(2ml/3T)^{1/2}$; (c) $\omega = (3T/ml)^{1/2}$.

5–15 (c) $(2 - \sqrt{2})^{1/2}\omega_0$, $\sqrt{2}\omega_0$, $(2 + \sqrt{2})^{1/2}\omega_0$, where $\omega_0 = (T/ml)^{1/2}$.

5–16 $\alpha = \cos^{-1}[1 - (\omega^2/2\omega_0^2)]$; $C = h/\sin[\alpha(N + 1)]$.

CHAPTER 6

6–1 (a) $\nu_1 = 10 \text{ sec}^{-1}$; (b) $\nu = 50, 100, 150$, etc., sec^{-1} (all integer multiples of 50 sec^{-1}).

6–2 $\nu_A = n\nu_1$ ($n = 1, 2, 3$), $\nu_1^2 = T/4ML$; $\nu_B = 0.84\nu_1$, $1.55\nu_1$, $2.04\nu_1$.

6–5 $\omega = \pi(T/LM)^{1/2}$.

6–6 (a) $\omega_n = [(2n - 1)\pi(Y/\rho)^{1/2}]/L$; (b) $\lambda_n = L/(n - \frac{1}{2})$; (c) $x = L(n - \frac{1}{2} \pm k)/(2n - 1)$ ($k = 0, \ldots, n - 1, \ldots$).

6–9 (b) $A_1 = 10 \mu$, $A_2 = 10 (1 - 1/\sqrt{2}) \simeq 3 \mu$.

6–10 (a) $\nu_n = nc/2L$; (b) (1) 21, (2) 15 cm.

6–11 (a) $(A^2n^2\pi^2T)/4L$; (b) $(A_1^2 + 9A_3^2)\pi^2T/4L$.

6–12 (a) $TL\{[1 + (2h/L)^2]^{1/2} - 1\} \simeq 2Th^2/L$; (b) every $2 (ML/T)^{1/2}$ sec.

6–14 $y(x) = \sum_{n=1}^{\infty} B_n \sin(n\pi x/L)$, where

(a) $B_n = \begin{cases} 8AL^2/(n\pi)^3 & n \text{ odd} \\ 0 & n \text{ even}; \end{cases}$

(b) $B_1 = A$, $B_n = 0$, if $n \neq 1$;

(c) $B_n = \begin{cases} \dfrac{(-1)^{(n+1)/2}4A}{\pi(n^2 - 4)} & n \text{ odd} \\ A/2 & n = 2 \\ 0 & n \text{ even}, n \neq 2. \end{cases}$

6–15 $y(x, t) = \sum_{n=1}^{\infty} C_n \sin(n\pi x/L)$, where

(a) $C_n = \begin{cases} 8AL^2 \cos(n\omega_1 t)/(n\pi)^3 & n \text{ odd} \\ 0 & n \text{ even}; \end{cases}$

(b) $C_n = \begin{cases} 8BL^2 \sin(n\omega_1 t)/n^4\pi^3\omega_1 & n \text{ odd} \\ 0 & n \text{ even} \end{cases}$

(ω_1 = angular frequency of lowest mode).

CHAPTER 7

7–2 (a) $A = 0.3 \text{ cm}$, $\lambda = 4 \text{ cm}$, $K = 0.25 \text{ cm}^{-1}$, $\nu = 25 \text{ sec}^{-1}$, $T = 0.04 \text{ sec}$, $v = 100 \text{ cm/sec}$; (b) $15\pi \text{ cm/sec}$.

7–3 $\xi = 0.003 \sin 2\pi[(x/600) + 5t)]$.

7–4 (a) $\frac{1}{3}$m; (b) $72°$.

7–5 (a) 22.4 m/sec; (b) 2.24 m; (c) $y(x, t) = 0.02 \sin(2.80x - 62.8t + 0.52)$.

7–6 (a) 10 m; (b) $y = A \sin(3\pi x/L) \cos(30\pi t)$.

7–7 $y = $ zero; $\partial y/\partial t \simeq 6 \text{ m/sec}$.

7–8 (a) $\nu = 1.5$ Hz;

(b) $\lambda = \dfrac{16}{16n - 1}$ m, $n = 1, 2, 3, \ldots$ for positive moving wave,

$\dfrac{16}{16n + 1}$ m, $n = 0, 1, 2, 3, \ldots$ for negative moving wave;

(c) $v = +8/5$ m/sec, etc., $v = -24$ m/sec, etc.;

(d) insufficient data.

7–12 (b) $v_y(\text{max}) \approx 4$ m/sec; (c) $T = 32$ N;

(d) $y(x, t) = 0.2 \sin 2\pi(8t + x/5)$.

7–13 (b) $v = u/2$, direction $= +x$;

(c) $\dfrac{\partial y}{\partial t}\bigg|_{t=0} = \dfrac{4b^3 u}{(b^2 + 4x^2)^2}\, x.$

7–16 (a) 8×10^{-4} sec; (c) $v_{\text{max}} = 12.5$ m/sec, during opening;

(d) $t = 1.2 \times 10^{-2}$ sec.

7–17 (a) $y(x, t) = 2A \cos\left(\dfrac{x}{2} - \dfrac{t}{2}\right) \times \sin\left(\dfrac{9}{2}x - \dfrac{19}{2}t\right)$;

(b) 1 m/sec; (c) 2π m.

7–18 (c) 50 cm.

7–20 (c) 28 m/sec $\simeq$ 63 miles/hr.

7–21 (a) $\lambda_n = 2l(N + 1)/n$; $\omega_n = 2\omega_0 \sin[n\pi/2(N + 1)]$;

(b) $v_p(n) = \omega_n \lambda_n/2\pi$,

$$v_g(n) = [2l\omega_0(N + 1)/\pi]\left\{\sin\left[\dfrac{(n + 1)\pi}{2(N + 1)}\right]\right.$$

$$\left. - \sin\left[\dfrac{n\pi}{2(N + 1)}\right]\right\}.$$

CHAPTER 8

8–1 $g_1/f_1 = 1, \frac{1}{3}, 0, -\frac{1}{3}, -1;\ f_2/f_1 = 2, \frac{4}{3}, 1, \frac{2}{3}, 0.$

8–3 $X = R$ for maximum dissipation.

8–4 $\omega = (LC)^{-1/2}$ for maximum dissipation, when L, C, R are given.

8–5 (a) 5.5×10^{-4}; (b) 1.1×10^{-3}.

8–7 (b) nearly 20 miles.

8–8 (a) total frequency drop $= 320$ Hz.

8–9 $T \simeq 900°$K.

8–12 (b) $v(\theta) - v'(\theta) \simeq v_0(u \cos \theta/v)^2$.

8–13 (c) speed $\simeq 0.1v = 110$ ft/sec; altitude $\simeq 1100$ to 1200 ft.

8–14 (b) All audio frequencies above about 1300 Hz; integer multiples of (approx.) 2200 Hz.

8–15 (a) 100; (b) $A/A_0 = \sin(100\pi N \sin \theta)/\sin(100\pi \sin \theta)$; (d) $\frac{1}{5}$.

8–16 (a) a distance much larger than 5 mm;

(b) for $d = 0.1$ mm, the ratio is roughly 0.44;

for $d = 0.05$ mm, about 0.05.

8–17 $I/I_{\text{max}} \geq 0.5$ for about 8 ft each side of maximum;

$I/I_{\text{max}} \geq 0.05$ for about 16 ft each side.

Index

Adiabatic compression, 59, 176
Air, elastic moduli of, 59, 178
   spring of, 57
Air columns, 57, 174
Amplitude (def.), 6
Anderson, O. L., 146, 147

Backus, J., 175, 303
Baker, B. B., 269
Ball, R., 252
Barker, J. R., 303
Barnes, R. B., 152
Barsley, M., 18
Barton, E. H., 87
Barton's pendulums, 87, 88, 92
Bate, A. E., 306
Beats, 22, 122, 215
Benade, A., 175, 303
Bergmann, L., 186
Bernoulli, D., 135, 168
Bernoulli, J., 135
Beyer, R. T., 244
Bickley, W. G., 306
Bishop, R. E. D., 3, 303
Bland, D. R., 304
Bloch, F., 110
Booker, H. G., 304
Bouasse, H., 77

Boyle, R., 57
Boyle's law, 58
Braddick, H. J. J., 304
Brillouin, L., 136, 304
Bulk modulus, 56, 176
Bunsen, R. W., 107

Characteristic impedance, 262
Chladni, E. F., 188
Chladni figures, 187, 188
Churinoff, G. J., 86
Coherence, 280
Complex exponentials, see Exponential,
   complex
Complex numbers, 10
Convective derivative, 227
Copson, E. T., 269
Coulson, C. A., 304
Coupled oscillators, 120, 124, 127, 136
   forced, 132
Crawford, F. S., 304
Critical damping, 70
Cronin, D. J., 296
Crystal lattice, 151
Cut-off, 234

Damped oscillations, 62
David, E. A., 7

313

Degeneracy, 184
Den Hartog, J. P., 304
Diffraction, single-slit, 288
Diffraction grating, 28, 284
Dispersion, 230
Doppler effect, 107, 274
Double slit, 280, 295

Eddington, A. S., 200
Elastic moduli, 46, 55, 58, 176, 210
Elasticity, 41, 45, 55, 57, 151, 176
Energy, in progressive wave, 237
   of harmonic oscillator, 42, 66
Energy densities, 238
Energy flow in wave, 241, 246, 295
Energy transport by wave, 241
Euler, L., 14
Euler's formula, 14
Exponential, complex, 13, 14
   use of, 21, 43, 64, 82
Exponential decay, 66

Feather, N., 304
Feynman, R. P., 14
Forced vibrations, 78, 83, 96, 168
Fourier, J. B., 5, 190
Fourier analysis, 168, 189, 191, 218
Fourier synthesis, 195, 222
Fourier's theorem, 5, 136, 190, 218
Frank, N. H., 145
Fraunhofer, J. von, 106
Fraunhofer diffraction, 293
Fraunhofer lines, 106
French, A. P., 296
Fresnel, J. A., 267
Frey, A. R., 305

Galilei, G., 166
Gas, elasticity of, 59, 176
Geneva, Lake of, 75
Gravity waves, 233
Group velocity, 233

Harmonic motion, see SHM
Harmonic oscillator, see Oscillator,
      harmonic
Helmholtz, H., 269
Herb, R. G., 108
Hooke, R., 2, 40, 41
Hooke's law, 40, 41, 42

Hudson, A. M., 185
Huygens, C., 267, 268
Huygens' principle, 267
   use of, 270, 275, 280

Impedance
   characteristic, 262
   electrical, 261
   mechanical, 259, 262
Impedance matching, 263
Ingard, K. U., 305
Interference, 281, 284, 294
Interference patterns, 282, 284, 285, 289, 296
Isothermal compression, 59

Jeans, J. H., 175, 305
Jenkins, F. A., 106
Josephs, J. J., 305

Kelvin, Lord, 118
King, J. G., 86, 296
Kinsler, L. E., 305
Kirchhoff, G., 107, 269

Lagrange, J. L., 190
Laplace, P. S. de, 245
Laplacian, 245
Leighton, R. B., 14
Lindsay, R. B., 244, 305
Lissajous, J. A., 35
Lissajous figures, 35, 36, 38, 45
Longitudinal oscillations, 57, 60, 144,
      170, 174
Longitudinal waves, 210, 264

Mach number, 278
Magnus, K., 305
Martin, W. T., 95
McCurdy, E., 229
McLachlan, N. W., 305
Miller, D. C., 162, 168, 215
Mode, see Normal modes
Moduli, elastic, 46, 48, 55; (tabulated), 47,
      56
Momentum of wave, 243
Morse, P. M., 305

Nodal lines, 281, 291
Normal frequencies, 126, 129, 141, 165

Normal modes, 119, 122, 129, 139
  of continuous string, 162
  degeneracy of, 184
  of loaded string, 139, 141, 147
  of membranes, 181
  orthogonality of, 196
  properties of, 141, 147
  of rods, 170
  spectrum of, 178
  superposition of, 124, 167
  of 3-dimensional system, 188
  and traveling waves, 202

Organ pipes, 175
Orthogonality, 195
  and normal modes, 196
Oscillations
  free, damped, 62
    undamped, 41, 48, 51, 54, 60
  longitudinal, 57, 60, 144, 149, 170
  transverse, 136, 139, 147, 162, 181
Oscillator
  anharmonic, 110
  damped, 63, 67
    energy of, 66
  forced
    damped, 83, 96
    power input, 96
    undamped, 78
  harmonic, 41, 43
    damped, 62
    energy of, 42, 66
  overdamped, 68
  torsional, 54
Oscillators, coupled, see Coupled oscillators
Overdamped oscillator, 68

Pain, H. J., 305
Pearson, J. M., 305
Pendulum
  rigid, 51
  simple, 49, 51
    driven, 81, 87
Pendulums, coupled, 121, 124
  driven, 132
Periodicity, 3, 6
Phase angle, 6, 80, 84
Phase lag, 80, 84, 89
Phase velocity, 233
Pierce, J. R., 7

Pohl, R. W., 267, 294, 297, 305
Polarization, 264
Power input to resonant system, 96, 98
Poynting, J. H., 36
Principal maxima, 287
Pulses, see Wave pulses
Purcell, E. M., 110
Pythagoras, 162

Q, 67, 89, 91
Quality factor, see Q

Radiation pressure, 243
Rayleigh, Lord (J. W. Strutt), 306
Reflection, 253
  partial, 256
Refraction, 270
Reissner, E., 95
Resonance, 77, 80, 89, 133, 169
  electrical, 102
  magnetic, 109
  nuclear, 108
  optical, 105
Resonance parameters, 91; (table), 105
Resonance width, 89, 98, 100, 101, 107,
    109, 110
Resonant frequency, 87, 91, 97, 98, 133, 169
Rigidity modulus, 55, 56
Ripple tank photographs, 267, 273, 277,
    282, 289, 292
Rods, speed of sound in, 210
  vibration of, 62, 170
Rosenfeld, J., 23, 25, 26, 38, 63, 88, 92,
    96, 122, 284
Rossi, B., 269
Rotating vectors, 7, 10
Rowland, H. A., 107
Runk, R. B., 146, 147

Sala, O., 108
Sands, M. L., 14
Sears, F. W., 255
Seiche, 74
Shear modulus, 55, 56
SHM, 5, 7, 15
  angular, 52, 54
  damped, 62
  of floating objects, 49
  geometric representation, 8, 44

of liquid column, 53
of pendulums, 51
SHM's, superposed
   different frequencies, 22
   equal frequency, 20, 27, 44
   parallel, 20, 22, 27, 37, 281, 285
   perpendicular, 29, 30, 35, 37
Shock waves, 277, 279
Simple harmonic motion, *see* SHM
Single slit, 288, 295
Slater, J. C., 145
Snell's laws, 270
Snowden, S. C., 108
Sonic boom, 279
Sound, 57
   speed of, 209; (table), 210
Spring, vibration of, 60
Standing waves (stationary waves), 164, 189
Starling, E. H., 4
Stephens, R. W. B., 306
Stoker, J. J., 306
Strain, 46
Straub, H., 4
Strength, tensile, 47
Stress, 46, 55
String, continuous
   forced vibration of, 168
   and Fourier analysis, 189, 193
   normal modes of, 162, 167, 189
   progressive waves on, 202, 207
String, loaded, 136, 147
   cut-off phenomena in, 234
Stull, J. L., 146, 147
Superposition, 19, 135
   of normal modes, 124, 135, 167, 189
   of progressive waves, 213, 232, 280
   of SHM's, 20, 22, 27, 29, 35, 37, 281, 285
   of wave pulses, 228
Sutton, O. G., 306

Talmud, 76
Taylor's theorem, 13
Temple, G., 306
Tensile strength, 47
Thompson, S. P., 268
Thomson, J. J., 36
Timoshenko, S., 306
Torsional oscillator, *see* Oscillator

Towne, D. H., 306
Transients, 92
Transverse waves, 204, 208, 213, 264
Tucker, W. S., 36

Undulation, female, 18

Van Bergeijk, W. A., 7
Vector diagrams, 286, 290
Velocity resonance, 97
Vinci, Leonardo da, 229

Waldron, R. A., 307
Waller, M., 187
Wave equations, 209, 228, 245, 246
Wave number, 214
Wave pulses, 216, 224
   Fourier analysis of, 219
   motion of, 223
   reflection of, 253, 256
   superposition of, 228
Waves, 201
   energy in, 237
   energy transport by, 201, 241, 246, 295
   longitudinal, 210, 264
   momentum flow in, 243
   and normal modes, 202
   progressive, 164, 202, 207, 230
   speed of, 164, 204, 210, 212, 233
   standing (stationary), 164, 189
   superposition of, 214, 232, 280
   transverse, 204, 208, 213, 264
   2- and 3-dimensional, 244, 265
White, H. E., 106
Width, *see* Resonance width
Wiener, N., 160
Wilberforce, L. R., 128
Wilberforce pendulum, 128
Wilson, H. A., Jr., 280
Wollaston, W. H., 107
Wood, A., 77, 307

Young, T., 46
Young's modulus, 46, 48, 56, 62, 151, 170, 210

Zemansky, M. W., 255